I0797311

The Surgeon, The Midwife, The Quack

The Surgeon, The Midwife, The Quack

How to stay alive in Renaissance England

ALANNA SKUSE

A Oneworld Book

First published by Oneworld Publications Ltd in 2025

A CIP record for this title is available from the British Library

ISBN 978-1-83643-077-3
eISBN 978-1-83643-078-0

Typeset by Geethik Technologies Pvt Ltd
Printed and bound in Great Britain by Clays Ltd, Elcograf S.p.A.

The authorised representative in the EEA is eucomply OÜ,
Pärnu mnt 139b–14, 11317 Tallinn, Estonia
(email: hello@eucompliancepartner.com / phone: +33757690241)

Oneworld Publications Ltd
10 Bloomsbury Street
London WC1B 3SR
England

Contents

Author's Note

Depending on who you ask, the Renaissance in England might start anywhere from about 1450 and last until about 1640, but in this book I've begun around 1580 and stretched the definition to cover the whole of the 1600s. The academic term for this period is 'the early modern' (or for archaeologists 'post-medieval'), but neither of those phrases offer the glitter that this part of history deserves.

I have resolved most of the seventeenth-century spelling and grammar idiosyncrasies, apart from those too wonderful to alter.

Dramatis Personae

Hannah Allen (c. 1638–?). Author, religious nonconformist, survivor of mental illness.

Robert Boyle (1627–1691). Anglo-Irish scientist, mathematician and experimenter. Member of the Royal Society. Friend or associate to *Robert Hooke, Margaret Cavendish, Richard Lower.*

Margaret Cavendish, Duchess of Newcastle (c. 1623–1673). Author, scientist and provocateur. Sometime patron to *Robert Boyle*, correspondent with (or reader of work by) *Sir Kenelm Digby*, *Henry More*, *Francis van Helmont.*

Elizabeth Cellier (first record c. 1668–last record c. 1688). 'The Popish Midwife'. Possibly associated with *Hugh Chamberlen the elder.*

The Chamberlen family (c. 1500–c. 1730). Family of physicians and man-midwives, including: Peter the elder (1560–1631); Peter the younger (1572–1626); Dr Peter (1601–1683) (son of Peter the younger); Hugh the elder (c. 1630–died after 1720) (son of Dr Peter); Hugh the younger (1664–1728) (son of Hugh the elder).

Anne Conway, Viscountess Conway and Killultagh (1631–1679). Philosopher. Friend to *Henry More*, patron of *Francis van Helmont* and *Valentine Greatrakes*.

John Cotta (1575–1627/8). Puritan physician and author.

Sarah Cowper (1644–1720). Noblewoman diarist.

Helkiah Crooke (1576–1648). Physician and anatomist, keeper of Bedlam Hospital.

Nicholas Culpeper (1616–1654). Astrologer, apothecary and physician. His widow Alice met with *John Ward*.

Sir Kenelm Digby (1603–1665). Courtier and scientist, member of the Royal Society. Associate of *Robert Boyle*, *Richard Lower*, *Christopher Wren*, correspondent with *Anne Conway*.

Elizabeth Freke (1641–1714). Memoirist and author of household physic recipes.

Valentine Greatrakes (1628–1692). Irish faith healer. Patronised by *Anne Conway*, associated with *Robert Boyle*.

William Harvey (1578–1657). Physician to Charles I, anatomist and scientist.

Francis Mercure van Helmont (1614–1698). Flemish alchemist and physician. Son of Jean Baptiste van Helmont. Patronised by *Anne Conway* and *Henry More*.

Richard Lower (c. 1631–1691). Scientist, anatomist and physician. Member of the Royal Society. Friend or associate to *Robert Boyle*, *Christopher Wren.*

Lady Grace Mildmay (c. 1552–1620). Memoirist and medical practitioner.

Henry More (1614–1687). Platonist philosopher. Teacher and later close friend to *Anne Conway*, patron to *Francis van Helmont.*

Elizabeth Okeover (1644–?). Medical practitioner and receipt book writer.

Ambroise Paré (1510–1590). Military surgeon, obstetrician and physician.

Samuel Pepys (1633–1703). Diarist and civil servant.

Hugh Ryder (dates unknown; active 1670s–1700s). Military surgeon and surgeon-in-ordinary to Charles II.

Jane Sharp (c. 1641–after 1671). Midwife and author.

Gaspare Tagliacozzi (1545–1599). Italian surgeon.

George Thomson (c. 1619–1676). Physician and campaigner against bloodletting.

Andreas Vesalius (1514–1564). Dutch anatomist and physician.

Thomas Vicary (c. 1490–1561). Surgeon and physician. First Master of the Company of Barber-Surgeons.

John Ward (1629–1681). Diarist, medical practitioner, vicar of Stratford-upon-Avon. Acquainted with *Richard Lower* and *Robert Boyle.*

John Westover (c. 1643–1706). Rural medical practitioner and hospice-keeper, specialising in care of the mentally ill.

John Woodall (1570–1643). Military surgeon, entrepreneur and diplomat.

Hannah Woolley (c. 1622–c. 1675). Medical practitioner, schoolmistress and author.

Christopher Wren (1632–1723). Scientist, architect, astronomer and mathematician. Founding member of the Royal Society. Associate of *Robert Boyle, Richard Lower.*

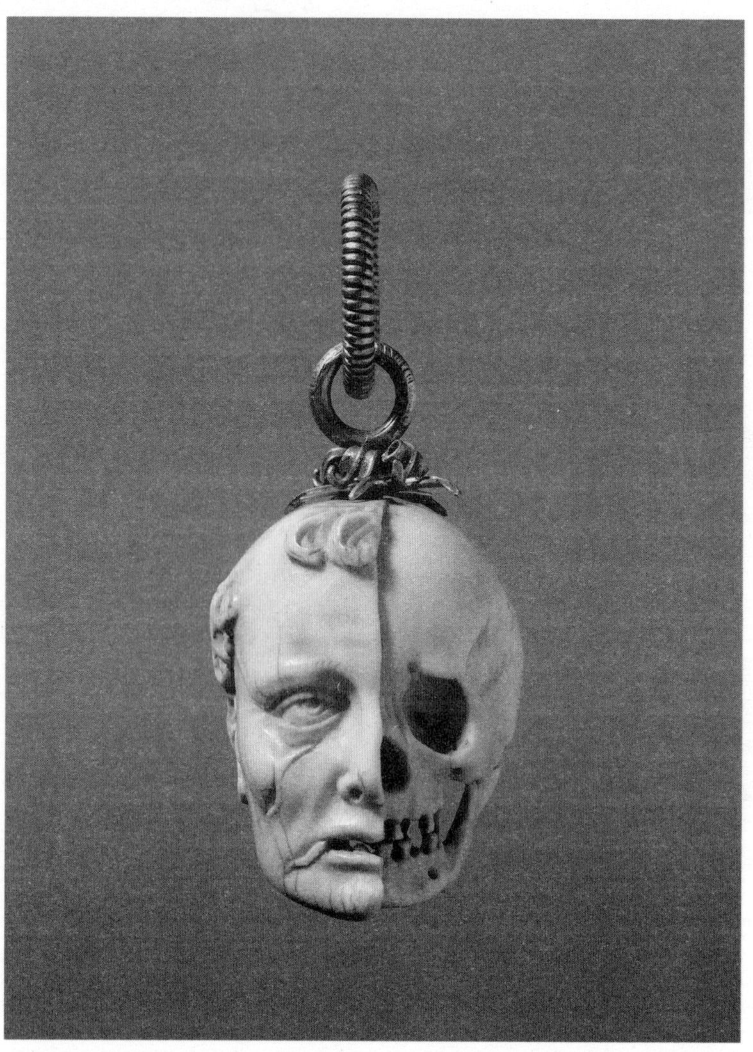

The terminal bead of an ivory rosary depicting the head of a man half decayed. Fashioned c. 1500–20 in what is now Germany, it was designed to remind the owner that death is always near. The Metropolitan Museum of Art, New York.

Prologue

Fetter Lane, London, sometime in the 1680s. In one of the houses crammed between the church of St Dunstan-in-the-West and the Inns of Chancery where court clerks ply their trade, a boy lies dying. A lawyer's son, he has probably lived here all his life, long enough to hear the Puritan Thomas Goodwin preach at the church up the street, long enough to whisper ghost stories to the merchants' boys at the end of the road where the gibbet used to stand. But a year ago, he was hit by one of the carts which rattle up and down the lane, and one dirty wound became eleven stinking fistulas in his leg, holes oozing fetid matter and breeding infection. Several doctors have visited; none has been able to help. As the leg lies folded beneath him, flesh slowly rotting away, his heel has become stuck to his buttock. Gangrene has eaten through ligaments and tendons like the rats that gnaw on ships' rigging, and his patella floats useless and unmoored. Desperate, the boy begs his latest visitor to lend him a knife, so he may cut off the leg and end his suffering.

There is just one last chance. Let down by their surgeon, his parents sought second, third and fourth opinions until they found Hugh Ryder, surgeon-in-ordinary to the king and one of the best men in the business. In his time aboard military vessels, Ryder treated men with brains exposed by flying shrapnel and stab wounds inches deep, but this is among the worst cases he has seen. He is moved by the family's tears and spurred on by professional curiosity. Still, he is careful of his reputation. Knowing the surgery could be fatal, he asks the local wardens to visit the child – they will be able to testify he'd been at death's door before Ryder's intervention. He waits one more day. The boy's mother is in labour again, and he fears that hearing her son's screams might kill her.

Finally, the dismemberment can begin. The boy stays silent as Ryder cuts through the rotten flesh with a curved knife, delicately slices the membrane covering the femur then saws at the bone with a foot-long serrated blade. He cries out only when hot irons are applied to the open wound. Though cauterisation is terrible, it is effective; the huge arteries of the thigh cannot be stitched fast enough to quell a deadly loss of blood. Nobody – not Ryder, his patient nor his family – really expects the child to live. Yet he does. The operation takes place on Monday, and when Ryder returns on Thursday, the stump is healing well, its edges pink and healthy. In six weeks, the boy is almost back to health, ready to begin a life on crutches.

***

The case of Hugh Ryder and his young patient shows how terrifying it could be to get sick or be injured in Renaissance England. In an era before antiseptics, anaesthetics or antibiotics, a simple broken leg could be a mortal wound – the tormented patient powerless to ease their pain, unable to access proper medical care, condemned to waste away over weeks and months.

But pause for a moment: might we see this tale in a different light? This patient's parents were able to find Ryder because, by the late seventeenth century, London contained more medical practitioners than ever before, rivalling Paris, Leiden and Padua in medical expertise. Ryder was able to attempt the operation because his professional training and years in the navy had given him vast experience operating on seemingly impossible cases. And by putting his experiences into print, Ryder paved the way for the next generation of surgeons to attempt their own bold cures. This is a story not only of suffering, but of compassion, curiosity and progress.

Today, we understand the Renaissance as a time of extraordinary vitality in the arts and sciences, a lightning flash that changed religious, political and cultural history. In a few centuries, the chokehold of the Catholic Church in Europe was released, the printing press made books widely available for the first time, the English Civil War upended the monarch's divine right to rule, and England transformed from a poor relation in European politics to a global power. Creative geniuses – William Shakespeare, John Milton, John Locke and Isaac Newton – transformed the literary and philosophical landscape. Yet in this age of illumination, one dark spot remains, a puzzle piece without which we cannot understand this vibrant, violent, changeable age. What did Renaissance people do when they got sick? When they were hurt, who did they send for? In an era so rich for the soul and imagination, what became of the body?

Our notions of sixteenth- and seventeenth-century medicine are of brutal, benighted practices, which offered few cures for people suffering desperately. Most medicines, we might assume, were ineffective or outright poisonous, and doctors couldn't identify the diseases they were treating anyway. People who hurt themselves faced having the injured limb hacked off by a 'sawbones' who was little better than a butcher, before succumbing to infection or blood loss, while the mentally ill were locked up and beaten into submission. In the bills of mortality that list causes of death in this period, the ailments can seem bizarrely quaint. The second week of April 1665 saw deaths from 'teeth', 'rising of the lights' and 'head-mould-shot', as well as five who simply died 'suddenly'.*[1]

In the sixteenth century, average life expectancy was just forty years, a number kept low by appalling rates of infant mortality (people who lived long enough to marry could expect to live into their fifties or sixties, or longer).[2] Of every hundred children born, around twenty would die before the age of five, eleven of those before reaching their first birthday.[3] Surgery was almost unthinkably awful, and because opening the torso of a living person was practically a death sentence, many simple complaints, like appendicitis, proved fatal, while illnesses like heart disease were poorly understood. Without antibiotic or antiviral medication, infectious diseases cut swathes through the population. During an outbreak of plague in 1665–6, up to 100,000 Londoners are thought to have died, one fifth of the city's inhabitants. Contemporary witnesses described bodies piled high on carts and church bells tolling continuously for the dead, with 'many poor sick people in the streets full of sores'.[4] It is little wonder that the well-off carried with them trinkets and jewellery adorned with skeletons and skulls to function as *memento mori*, reminders that death was

* 'Teeth' referred to children who died during teething, often of fevers. 'Rising of the lights' was an obstruction of breathing, often associated with croup. 'Head-mould-shot' meant injury to the bones of the skull, or a congenital skull malformation.

coming. When life was so precarious, any sensible person kept themselves spiritually prepared for the hereafter.

Yet people in this period did not simply fall ill and wait to die, any more than we do today. When we look behind the grim statistics to the lives of individuals, we find patients striving to survive and medical practitioners working tirelessly to help them. Over the course of a century and a half, these individual stories add up to an extraordinary expansion of medical knowledge and the medical marketplace, giving rise to a booming industry populated by innovators and trailblazers as well as charlatans and crooks. When Henry VIII was born in 1491, he entered a world in which medicine remained largely in the hands of monks, nuns and local wisewomen, whose methods relied heavily on charms and prayers. Barbers and physicians – trained, but entirely unregulated – were spread thinly among larger towns and cities, tending to the rich. By 1700, ten monarchs (and one regicide) later, healthcare was practised by a cast of characters from all walks of life, including physicians, surgeons, midwives, nurses, apothecaries, bone-setters, tooth-pullers and creators of all manner of home remedies. With ingenuity and fortitude, they calmed frenzied minds, treated embarrassing rashes, helped women through labour and tended to babies in the perilous first months of life. They might even give new faces to those disfigured by injury or disease, make the lame walk again and help the infertile to beget children.

This book tells the stories of those exciting, frustrating, caring and contradictory practices and practitioners which touched the life of every Renaissance person, and which continue to shape how we think about health and medicine to this day. Each chapter is dedicated to a different group, but as always in histories, some groups are represented more than others. Physicians, for instance, made up a relatively small proportion of all the people practising medicine in this period, but, being highly educated and literate, they left a lot of records of their activities. On the other hand, not much was written by or about travelling medicine-sellers, so they are grouped in with other important but ephemeral figures, such as bone-setters and tooth-drawers. The most influential and elusive actors in the medical marketplace were patients, whose needs, means, hopes, fears and testimonies determined which practices and practitioners thrived, and which faded into obscurity. Where they can be glimpsed on the margins, I have tried to bring them into the spotlight.

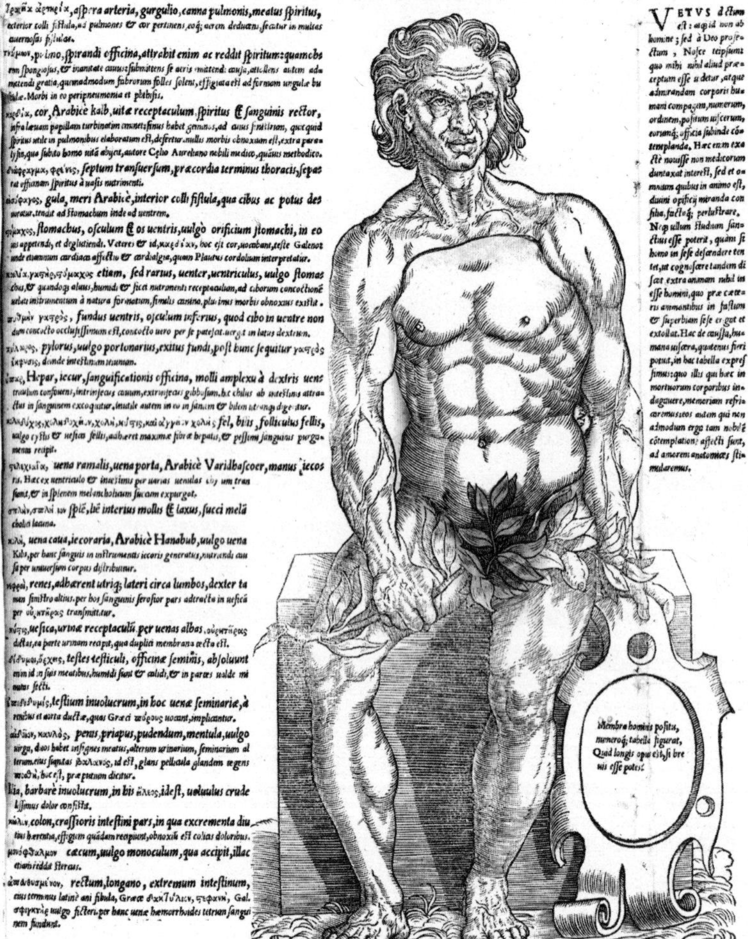

A page from an Italian ‘anatomical fugitive sheet’, 1539. The leaves mark the edge of a flap which can be lifted to reveal the internal organs. The Wellcome Collection.

INTRODUCTION

Know Thyself

What did Renaissance medics know about the human body? Well, both more and less than we might think. Medical practitioners of the sixteenth and seventeenth centuries had no way of looking inside their patients besides cutting them up and, until microscopes became popular in the 1650s, no way of viewing cells or microbes. They knew nerves were vital to sense and motion, but were unsure how, and had to posit elaborate yet vague theories of 'vital spirits' and 'volatile blood' to fill the gap. Some of their accepted precepts – such as the existence of a special vessel in women that turned blood into milk – were based on little more than imaginative speculation. Nonetheless, they were also keen observers, who painstakingly mapped the veins and arteries, described the structure of the eyes and could discern differences between a malarial fever and influenza, 'falling sickness' (epilepsy) and fainting, benign growths and cancerous tumours.

More importantly, each generation of medical practitioners knew *more* than their predecessors, because, thanks to the continual refinement of the printing press, they finally had access to one another's observations and opinions via cheap, readily available books published in *English*, rather than Latin. This development would later prove crucial for the growth of medicine as an industry; it drew medics of all kinds into a web of discourse and debate, allowing them to boast, squabble and advertise their wares as never before. Chiefly, however, books informed both practitioners and patients about two areas of knowledge and investigation that defined medicine in this era. The first was the system of the humours, known as 'humoralism' or 'Galenism'. The second was the nascent field of human anatomy.

Restless bodies: the four humours

Four Humours reign within our bodies wholly,
And these compared to four Elements,
The Sanguine, Choler, Phlegm, and Melancholy.
The latter two are heavy, dull of sense,
The t'other are more Jovial, quick and Jolly,
And may be likened thus (without offence)
Like air both warm and moist, is Sanguine clear,
Like fire doth Choler hot and dry appear,
Like water, cold and moist is Phlegmatique
The Melancholy cold, dry earth is like.[1]

This rather ropy poem, by an author known only as 'Joannes', appeared in a sixteenth-century book called *The Englishman's Doctor*, which delivered all its recommendations for a healthy diet and lifestyle in verse. Presumably, Joannes realised too late that he didn't have a rhyme for 'phlegmatic' and tried to torture the word into a form that would rhyme with 'like' instead. Nevertheless, the poem neatly sums up a unique and surprisingly intuitive model of physiology, which can help us understand why sixteenth- and seventeenth-century physicians and patients embraced medicines that seem useless or downright harmful today. During the

Renaissance, the body was imagined as a teeming cauldron of fluids known as 'humours', which clashed and coexisted in the blood. Each of the four humours – melancholy, choler, phlegm and blood – had its own qualities, based on its position in the quadrant of heat and moisture and linked to the elements of earth, air, fire and water. Together, they are key to understanding the logic of Renaissance medicine, and the way that ideas of sickness and wellness permeated every aspect of everyday life.

The humours were not a new idea. They'd been around for millennia, first written down in ancient Greece by Hippocrates of Kos and his followers, and then by Claudius Galenus, known as Galen of Pergamon (126–219 CE). Though Hippocrates is now known as the father of medicine, for Renaissance readers Galen was the true star.[2] Born in Pergamon, in modern-day Turkey, he grew up in a town renowned for its fine temples and a library said to be second only to the Great Library of Alexandria. According to legend, Galen was always fated to be a physician; his father received a vision urging him to send his son to medical school at the local healing temple (an *asclepeion*, named for the Greek demigod of healing, Asclepius). However, the young prodigy's true talents did not emerge until he left the temple to travel the Mediterranean. In Greece, Crete, Egypt and Cyprus, he observed outbreaks of disease and watched local healers at work. By the time he reached the splendour of Rome, he was committed to a new style of medicine that rejected the old tools of mystical divination and sought to understand the body's workings from the inside out. He first conducted vivisections and anatomies on primates, but their faces were too human for his comfort, and he soon switched to pigs (an understandable choice, but one that led to some rather odd conclusions about human bodies).

Many of Galen's anatomical findings would be overturned in the Renaissance, but they marked the first systematic attempt by a European author to understand mammalian physiology.

When medieval and Renaissance physicians turned to the classical world for new medical ideas, it was Galen's work they found most compelling. He seemed to epitomise everything to be revered about ancient Greece and Rome, bygone eras more civilised than the current age. Not just a medical practitioner, he was an elegant (and prolific) writer and philosopher who synthesised the ideas of great thinkers like Aristotle and Plato. Most significantly, though, his theory of the four humours offered something for everybody. At its root, it was simple enough for anyone to understand, but its details were complex enough that physicians could study and argue over its nuances for as long as they wanted. It suggested that certain dispositions of the body were inevitable, so it did not blame people for their frailties, yet it also left room for those dispositions to be altered by medicine, diet and other environmental factors. Last but not least, it was a complete system, which encompassed mental as well as physical health; the humours shaped aspects of a person's personality as well as their susceptibility to illness. The seventeenth century brought new ideas and new physiological models, but for many physicians and their patients, humoralism would always be the biggest show in town.

Plenty of Renaissance textbooks talked about the humours, but among the most comprehensive was a slim volume by the Northamptonshire physician James Manning, catchily titled *Complexion's Castle*. In 1604, a year after the coronation of James I, the book explained to the public in plain terms their humoral 'complexions' and what they ought to do about them. Knowing one's own body, Manning argued, was essential to both mental and physical health. Luckily, the humoral system was quite intuitive. When you knew the basic qualities of the four humours, their effects – and their remedies – made perfect sense.

First on Manning's list was *blood*, the 'sanguine' humour, warm and wet. Confusingly, this humour was different to the 'nutritive blood', the red liquid within which all the humours flowed. Those in whom the sanguine humour dominated – often young men – were generally hale and hearty. Their faces were 'white and ruddy' with large eyes and rosy cheeks, their hair red or blond and their beard 'large and comely' (no coincidence that this favourable complexion neatly described the features of the Tudor lineage). Full of moisture, they sweated and urinated profusely and enjoyed 'perfect' digestion. They might suffer from a stitch now and again, the product of their love for culinary indulgence, but were less subject to infections than other complexions.[3] What was more, they were liberal and jolly, if somewhat prone to get themselves in trouble with their love of wine, women and song.

Next to blood was *phlegm*, a cold and wet humour. Phlegmatic people were typically pale, flabby and slow to act and react. Manning explained how they appeared 'sleepy, dull' with 'plain hair', largely hairless bodies and 'the feet stinking, especially after motion'. Phlegmatic people found themselves 'subject to pain in the head, loins, knees, privities [genitals], arms, and feet, by phlegmatic humours taking cold'. They were prone to respiratory complaints and dropsy, a build-up of fluid in the tissue which could reach amazing proportions. In their characters, they were 'fearfull', cowardly and 'timorous', but slow to anger and quick to forget their grievances.

Though being constitutionally cold and wet was hazardous to one's health, phlegm's opposite humour, *choler* (also called 'yellow bile'), was even worse. Hot and dry, this humour attracted infections to itself, sucking up moisture from the air, and with it, pestilence. Choleric people could be spotted by their lean, sallow appearance, dark hair, darting eyes and sharp voice. They were vivacious, but also quick-tempered, sometimes covetous, and possessing a lusty sexual appetite. The physician might also notice their 'high coloured' – that is, very yellow – urine, and their strong pulse. Even more than other complexions, choleric people needed to pay heed to their humours if they were to avoid gout, fever and fits. Anything that heated the body still more, such as exercise, sex or certain foods, presented a real danger of burning up one's insides, with dire consequences.

Finally, Manning described the cold and dry humour of *melancholy*. Translating as 'black bile', it was a slow, thick humour like the dregs of wine, which crept through the veins and chilled all it touched. It could stagnate in the body's crevices, causing hard tumours. When it met with the fiery choleric humour, the blockages it caused were more malign, turning to cancers which overtook the body like a ravening wolf. Melancholy humours were betrayed by a lean body, 'plain' brown or grey hair, and skin that was 'hard' and pale. The melancholy person moved slowly, often finding themselves short of breath or struck with 'heaviness'. Unsurprisingly, their character was sad and serious, 'long in anger or dislike' and 'seldom laughing'.[4] In serious cases, this temperamental tendency could develop into the disorder of melancholy, a complaint similar to clinical depression, of which the deadliest side-effect was suicide.

As Manning's catalogue laid out, the humours were more than just an element of physiology. In these four fluids also dwelt the seeds of a person's personality; having 'a sense of humour' or being 'good-humoured' comes directly from these 1,800-year-old ideas. *The Englishman's Doctor* explained, 'Complexions cannot breed virtue or vice / Yet they may unto both give inclination.'[5] At every moment, the humours influenced the emotions, and the emotions moved the humours, but the terms of this perpetual negotiation were a mystery. Thomas Wright was the foremost scholar in this area, yet even he admitted, 'As the motions of our passions are hid from our eyes, so they are hard to be perceived'.[6] Still, he could speculate based on what he knew of people and their habits.

When a person perceived or remembered something, Wright argued, mysterious 'spirits' flocked from the brain to the heart, 'signifying' the thought to that organ. The heart then jumped into action and, to help the body to gain or avoid the object of the thought, 'draweth other humours to help him'. In pleasure, 'pure spirits' abounded, 'in pain and sadness, much melancholy blood, in ire, blood and choler'. Organs all around the body contributed humours to the heart, like obedient nobles paying tribute to their king. But what the heart could gather depended on what humours were dominant in the body, for 'if one abound more with one humour than another, he sendeth more fuel to nourish the passion, and so it continueth the longer, and the stronger'. A temperamentally melancholy individual, for example, might find that they had little ability to rouse themselves to anger even in the face of great provocation, but could be desperately sad for almost any reason.[7] These beliefs shaped not only how people perceived themselves, but how they presented themselves to others. For instance, many writers claimed to abound in melancholy humours. A portrait of the great poet and preacher John Donne shows him to be the perfect melancholic: slender, with abundant dark hair and a doleful gaze. He had a considerable flair for the dramatic and, in 1631, when he felt (accurately, as it turned out) that he was nearing death, he commissioned a statue of himself which still stands in St Paul's Cathedral, one of the few monuments to survive the Great Fire in 1666. In it, he is wrapped in a funeral shroud, standing on an urn containing his own ashes.

Like warring gods, the humours jostled for supremacy within the human body. To balance the humours was an impossible task – as soon as equilibrium was achieved, it was lost again. Just which humour would dominate any given body depended on a raft of factors. There were some outside of anyone's control, like nationality, age and gender. Part of why the Italians and Spaniards were supposedly so hot-tempered was because of their abundance of choleric and sanguine humours, exacerbated by warm weather, too much sex and all the garlic they ate (think of Shakespeare's perpetually warring Montagues and Capulets). Children were born with an abundance of heat and moisture, which gradually subsided throughout their lives until, by old age, they dried up like shrivelled fruit. Men were naturally hotter throughout their lives, which explained why they had to keep their genitals on the outside to avoid overheating them. Women, by contrast, were cooler and wetter, which accounted for their softer flesh and (as men would have it) slower brains.

As important as these facts of nature, however, were the so-called 'non-naturals', the environmental and circumstantial factors that made one person's humours different from the next. Non-naturals included the air and weather, exercise, excretions (including sexual), sleep, diet and 'mental affections' (thoughts and emotions). In fact, virtually every activity and substance could be categorised as healthy or unhealthy. Farting, for instance, was a medical necessity, for 'Great harms have grown, and maladies exceeding, / By keeping in a little blast of wind'.[8] Washing one's face and teeth was hygienic, but excessive bathing opened the pores and allowed diseases to enter. Exercise such as walking and riding was wholesome, and being outdoors was generally a good thing, as long as one could avoid catching a chill – or worse, breathing in 'putrefactive air'. Though germ theory would not become established until the nineteenth century, even small-time physicians observed that somehow, people tended to get ill at the same times in the same places. The signs of bad air ranged from stagnant water, to overabundant frogs, to shooting stars. Physicians and patients alike were disturbed by the 'portent' of Kepler's supernova, which appeared in 1604, and by Halley's comet, which blazed through the sky in 1531, 1607 and 1682.

The most talked-about of the non-naturals, however, were undoubtedly food and drink, topics which took up hundreds of pages in physicians' textbooks. Obsession with nutrition may seem like a twenty-first-century phenomenon, but four hundred years ago, patients were agonising over not only the best foods to eat, but how to balance them optimally and when to eat them (spices and wine in winter; lighter foods in summer). Broadly speaking, strongly flavoured savoury foods tended to heat the blood, and therefore promote hot and dry choleric humours. Onions, garlic, cheese, most red meat, offal, strong wine, spices and salt were all to be taken judiciously, for they had power to alter the body's constitution. Such 'hot' foods were best suited to teenage boys, who needed them for rapid growth, and old men, who could use them to redress their frigid humours. Freshwater fish, poultry, green vegetables, eggs and milk were safer alternatives, which moistened and cooled raging tempers. Sugar and honey were thought to be extremely nourishing, leading to widespread tooth decay among those who could afford these delicacies on a regular basis. Wine – especially white wine – and water were deemed relatively healthy drinks (noblewoman Sarah Cowper took care to drink a half pint of wine each day to counteract the cooling humours of her middle-aged body). Beer was of variable benefit, and cider was potentially harmful, not because of its alcohol content but because it was made of fruit, which was in general harmful or 'noisome to man', to be eaten very sparingly.[9] For the sick, broth or 'caudle', a warm drink of ale or wine mixed with spices and eggs, was a wholesome treat.

It was a neat theory, provided one had the financial means and personal discipline to follow a humour-friendly diet. However, then as now, patients repeatedly infuriated their doctors by indulging in their favourite dishes regardless of the consequences. 'Patients are like children,' complained one surgeon, 'still desiring such things which are offensive and hurtful'.[10] A prime example was the seventeenth-century diarist Samuel Pepys, a man as casual about his diet as he was about his marital fidelity. On numerous occasions in his diaries, he records enjoying large quantities of cider, often before sexually harassing one of his long-suffering maids. Nonetheless, when on 9 February 1663 he found himself confined to bed with stomach pains and fever, he was quick to blame the large quantity of 'Dantzig-gherkins' he'd eaten the day before. As a remedy, he took Venice-treacle, also known as theriac, a popular recipe containing, among other ingredients, opium and snake venom.

The idea that diet, exercise and environment have an impact on wellbeing might seem tediously obvious. But for Renaissance people, this was part of a worldview in which physical and mental health were one and the same, so factors that adjusted the proportion of humours actually altered the self in the most fundamental sense. Cutting back on red meat didn't just quell a fever; it could change your appearance, calm your emotions and help you to think in a more measured way. What was more, the exchange worked in both directions; changing your thoughts could help change your emotions and in turn, your body. In 1670, for instance, an anonymous author stated confidently, 'We are assured by many observations, that cancers sometimes have proceeded from the stopping of the monthly visits [i.e., menstruation], by frights, violent grief, and by letting of blood unseasonably.'[11] Emotional and physical experiences were here seen as much the same thing, producing the same effect.

The humoral model was extremely popular and enduring, surviving in one form or another for over 1,500 years. It flourished in the Renaissance because it offered some chance to understand and shape one's health even in an era when death was all around. Yes, a free choice of diet was reserved for the lucky few, and a change of air was a privilege available only to those rich enough to keep both town and country residences. But most people of the 'middling sort' could link their moods and maladies to their daily routines and act accordingly.* Arguably, there was a moral duty to guard one's own health, and James Manning warned his readers in typically fulsome style that the man (or woman) who neglected their own humoral wellbeing was only one step away from that

* There was not a defined middle class in this period, but there was an increasing number of skilled labourers, artisans, civil servants, merchants and clergymen, with moderate freedom and purchasing power, whose families were neither aristocratic nor poor. These are known to historians as the 'middling sort'.

most reprehensible creature, the suicide. 'The soul crieth unto thee to correct bad humours,' he warned, 'for when she would be gentle, mild, and patient, the excess of choler, constrayneth to rage and revenge, when she would watch and pray, the excess of phlegm, causeth sleepiness, and dullness: likewise the excess of other humours, or the confounding of humours, worketh effects more unkindly against the soul.'[12] In Manning's book, healthiness was next to godliness.

Humoral theory would start to be challenged in the middle of the seventeenth century, gradually falling out of favour in the 1700s. As scientists peered through increasingly powerful microscopes and observed ever tinier worlds, they began to rethink their model of the body. Nonetheless, humoralism was vital to the medical revolution of the sixteenth and seventeenth centuries, because it encouraged people to believe that their health was in their own hands. It is often thought that Renaissance people simply accepted their illnesses as divinely ordained. The reality was quite different: though the power of life and death ultimately rested with God, God helped those who helped themselves.

Looking inward: anatomy

Strangely enough, the vision of the humours on which so much of early modern medicine depended did not require a particularly intimate knowledge of the human body. Renaissance men and women were, generally speaking, more familiar with blood and guts than we are today. As we will see, many ill people were cared for at home, and the better-off would always choose to have surgery in their own rooms rather than take their chances in one of the filthy hospitals reserved for society's poorest. Likewise, childbirth took place at home, surrounded by female relations and 'gossips'. People of middling status or lower would have seen animals being butchered and their carcasses being prepared, or attended bear-baitings and cock-fights. A good day's entertainment for a young apprentice in London would have been a trip to Southwark, seeing two cockerels tear into each other with sharpened talons, before watching a tragedy unfold in the nearby Globe Theatre, maybe rounded off by visiting one of the city's sex workers – also known as 'Winchester geese' – who frequented that neighbourhood. On the way home, they might have passed by the traitors' heads mounted on pikes at the end of London Bridge. Dipped in tar, pecked at by birds, they were a grim warning against rebellion, and there was plenty of rebellion about. When he visited the city in 1596, the German lawyer Paul Hentzner counted over thirty heads.[13]

Before the Renaissance, this was as close as most people, even doctors, got to investigating the inside of a human body. When we look at representations of human anatomy from the Middle Ages, they are very beautiful...but *very* strange. Anatomised figures squat with feet spread across the page, their organs represented separately, like islands on a map. They are roughly in the right places, but they give no sense of how organs crowd and press into each other; how the lungs embrace the beating heart and the intestines coil in the trunk of the body. The major blood vessels are sketched onto the arms and legs, but there are no delicately dissected hands and arms with each tendon connected to its corresponding bone, or brains cradled by protective layers of membranous tissue. Instead, medieval Christianity treasured the body principally for the clues it might provide to divine mysteries.

In the late thirteenth century, Chiara of Montefalco, an Italian nun who provided medical care to the poor in her community, claimed that God dwelled in her heart. Her holy sisters took her at her word, and when she died in 1308, they promptly opened her corpse to look inside. Sure enough, they reported, Chiara's heart contained the image of a cross upon its walls, and her gall bladder housed three stones, symbolising the Holy Trinity.[14]

Compare this with Rembrandt's *The Anatomy Lesson of Dr Nicolaes Tulp*, painted in 1632. Here, a group of men in black gowns and starched white ruffs encircle the pale naked form of an unnamed cadaver. The picture has sparked many imitations and parodies: would-be tattooists looking at an inked body, a controversial image of Nelson Mandela under the anatomist's knife, or – more surreally – the assembled Muppets peering disconsolately at the prone figure of Kermit the Frog. However, in the original, the lesson Dr Tulp is teaching seems to be correct. He shows his assembled students the muscles of the arm, which Rembrandt has reproduced in faithful and astonishingly accurate detail, from the flexor digitorum superficialis to the tendon flexor pollicis longus.[15] If God is in this dissecting room, He is not making marks on the heart, but channelled through the figure of the knowledgeable doctor.

A clue to how this transformation came about can be seen at the edge of Rembrandt's painting. Visible in the lower right-hand corner is a large open textbook, thought to be Andreas Vesalius's 1543 *De humani corporis fabrica* ('Fabric of the Human Body'). Vesalius's seven-volume set, completed over thirty years in the early sixteenth century, was the definitive break from old models of anatomy by Galen, and the gold standard for a new kind of anatomical study based on observing the body as it really

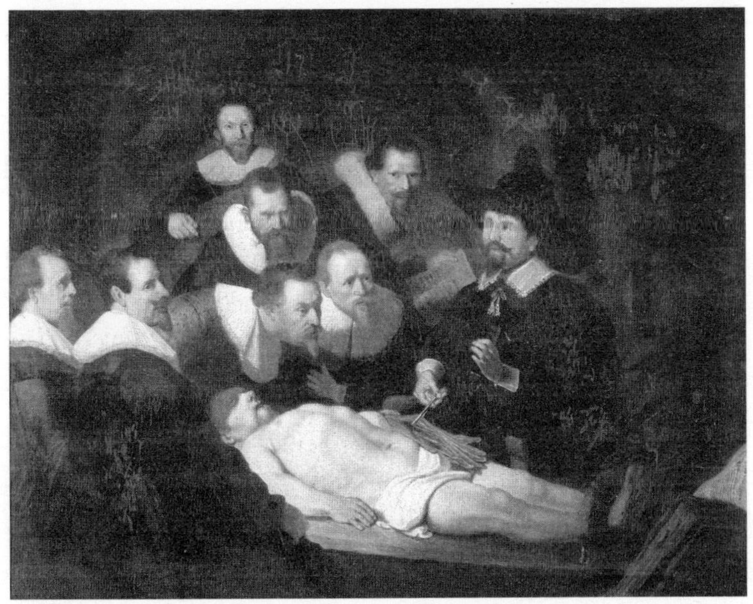

The anatomy of Dr Nicolaes Tulp. Oil painting after Rembrandt van Rijn. The Wellcome Collection.

was, not as ancient scholars supposed it to be. One idea Vesalius debunked was the notion that men had one fewer rib than women, supposedly because God had taken a rib from Adam in order to make Eve.

The *Fabrica* was a gorgeous and covetable book, capitalising on the latest advancements in drawing, woodcuts and printing. Its most lavish edition was presented to the Holy Roman Emperor Charles V, bound in purple silk with the illustrations coloured by hand. It helped the ambitious author secure a position as the emperor's personal physician. The fact that Vesalius correctly deduced that the devoutly Catholic Charles V would appreciate such a gift should give us a clue that the sixteenth-century Church did not prohibit human dissection, as has often been claimed. In fact, the Church never completely banned dissection of human corpses, even in the medieval period. Instead, the forces at work were more prosaic: people did not generally want to have their relatives cut up after death. The Bible said that the righteous would be resurrected *in their bodies* at the Last Judgement, and though it was taught that God could reconstitute a body that had lost parts here and there, most did not want to run the risk.* Therefore, anatomists were left with what they could scavenge from poor-hospitals or the scaffold. As towns and cities expanded, more bodies became available. As universities grew, especially in Italy, there were also more students who wanted to see anatomies, and more enterprising characters like Vesalius (whose name translates to 'weasel'), who realised that anatomies made good science *and* good business.

Anatomy theatres were aptly named; they combined education with spectacle, science with art. Famous examples like those in Padua, Leiden and, later, London, attracted rowdy crowds of medical students and curious laymen, eager to see nature's secrets laid bare. They shared a distinctive tried-and-true design. Concentric circles of bench seats rose steeply upward, surrounding the dissection table. Popular theatres could seat several hundred people. For extra flourish, some theatres included natural curiosities as part of their offering. In 1683, Jacobus Voorn

* John Donne, the melancholic poet-preacher with the statue in St Paul's was obsessed with this theme, and wrote swooping poems and sermons imagining how God might reconstitute a soldier's body which had lost an arm in one ocean and a leg in another. Cautious relatives might also recall how Jesus was said to have appeared to Saint Thomas after the Resurrection, inviting him to touch the (very much intact) wounds made during the crucifixion.

published a list of the 'chiefest rarities' to be seen at the theatre in Leiden. In the entryway alone were the head and feet of an elephant, the heads of an elk and a rhinoceros, the skeleton of a small whale removed from a larger whale's belly and a pair of skis – or as Voorn knew them, 'stilts with which the Norwegians, Laplanders and Finlanders run down high snowy mountains, with almost an incredible pace'.[16] In the theatre itself were many more animal skeletons, plus those of several human criminals, while various cases and cupboards contained further wonders, from 'a piece of rhubarb grown in the shape of a dogs head' to yet more skeletons, amber, reptile skins, and 'the mummie of an Egyptian prince above 1300 years [dead]'.[17] These would be exhibited in the theatre during the summer months, when bodies rotted too quickly and smelled too bad to allow for anatomies to take place.

At the centre of the anatomy theatre was the intimate encounter between corpse and anatomist. In the fifteenth and early sixteenth centuries, this had been conducted at a chaste distance. The physician-lecturer sat high up in the theatre reading from Galen's anatomical textbook, while a barber-surgeon did the dirty work of opening up the body. A third person might stand by with a long stick pointing out the relevant parts; the theatre was lit by candles, and in the winter gloom it could be difficult to differentiate between closely packed organs.

Vesalius was one of the first physicians to descend from his perch and get up close and personal with the cadaver. The frontispiece of the *Fabrica* shows him looking towards the reader as he delves into the body of a woman. Spectators are crowded around him, pointing, clamouring for a better look, some even jostling for position under the anatomising table. There's a monkey on one man's shoulder, and a dog in the corner hoping for a bite to eat.

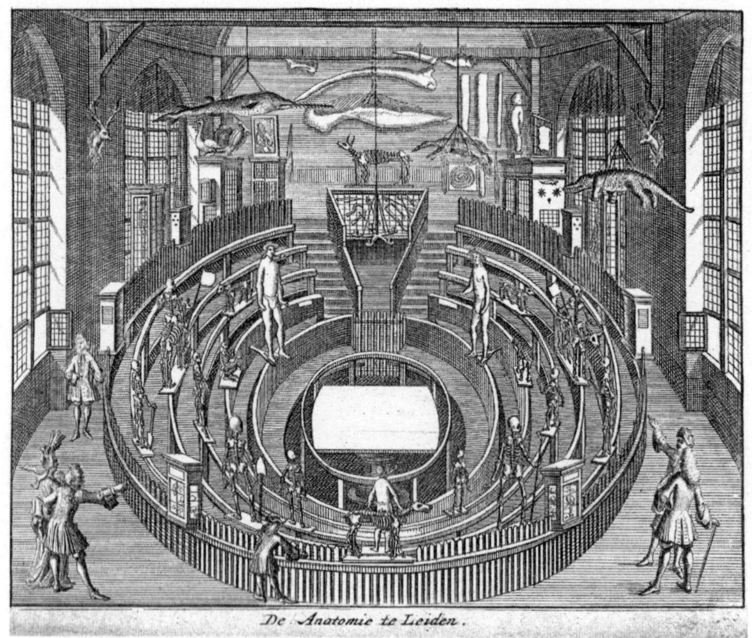

Leiden Anatomy Theatre, showing some of the exhibits housed there in the summer months. Line engraving. The Wellcome Collection.

The anatomist looks out placidly through the chaos, as he calmly organises the confusion of the body's bloody interior into its logical, explicable parts. Each anatomy followed a similar pattern, designed to make the most of corpses which were hard to come by. First to be examined were the viscera, the soft organs in the centre of the body which would rot the fastest. Then followed the rest of the trunk, including the heart, and the brain. The sinewy limbs could be perused at greater leisure, and another portrait of Vesalius shows him dissecting an arm in the comfort of his study.

The *Fabrica*, and the rising popularity of anatomies in general, helped to change medical practice. Some changes were practical and immediate. For example, Vesalius accurately depicted

Frontispiece from Andreas Vesalius's *De humani corporis fabrica libri septem*. The Wellcome Collection.

the meninges (protective outer membranes) of the brain, which meant that surgeons treating skull fractures could carefully lift embedded bone away from the delicate brain and, if necessary, stitch up the dura mater beneath. More importantly, though, anatomies wrought a cultural shift, emboldening scholars to believe that the most obscure of nature's mysteries could be subjected to scientific study. Instead of a locked vault, the body became an unread book, and physicians were the translators. Through the seventeenth century, successive generations of medical practitioners would publish new findings on the brain, the nerves, the workings of the heart and the reproductive system. The *Fabrica* became *the* reference text for would-be physicians and surgeons, even though fewer than five hundred copies were printed, and none in English.[18] In 1638, an even bigger and better version of the famous anatomy theatre at Padua was built for the Company of Barber-Surgeons on Monkwell Street, London.[19]

Anatomy sought to bring human physiology into the light of understanding, but there was also a darker side to the endeavour. Anatomists were desperate for corpses, especially female ones, since women were less likely to die on the scaffold and be offered up for examination. The ultimate prize was a pregnant woman's body, a rare find indeed. In the 1543 edition of the *Fabrica*, Vesalius happily admitted that many of his students broke into tombs to obtain bodies to bring to his classes, stealing corpses of children as well as adults.[20] By the 1555 edition, the most graphic of these stories had disappeared. Vesalius had caught up with public opinion that abhorred such goings-on, but the damage was already done. Rumours began to fly, not only that anatomists were habitual grave-robbers, but that when their attempts at robbery were foiled, they would turn to vivisection.

The first of these accusations was probably true, especially on the continent, where private anatomies could make large sums for all those involved (all, that is, except for the main attraction). The second claim – that anatomists vivisected living people – was harder to pin down. Throughout the Renaissance, scientists conducted horribly cruel experiments on animals to gratify their curiosity. They routinely cut up live dogs, monkeys and pigs. Sometimes they rearranged their inner workings and stitched them back up, or more often they let the poor creature die. Though this period was often unsentimental about animals, people had pet dogs and monkeys, and they wondered at the men who could ignore such pitiful cries. Then there was their ghoulish enthusiasm for getting corpses as fresh as possible. In 1565, the French diplomat Hubert Languet wrote a letter to his physician friend Casper Peucer, gossiping about why Vesalius had needed to leave Spain in a hurry the previous year. Languet claimed that 'A Spanish nobleman had been entrusted to his care, but when Vesalius believed him to have died…he sought permission from the relatives of the dead man to open the body; having obtained such permission, when he opened the chest he found the heart *still beating*.'[21] Elsewhere, Gabriele Falloppio, the man who gave his name to the fallopian tubes, cut out the middleman by persuading the Grand Duke of Tuscany to give a condemned criminal directly to him to kill and then dissect. He gave the condemned man two drams of opium, but he did not die. They made a deal – he would give him two more, and if he lived, Falloppio would intervene with the duke to secure a pardon. The man died, and Falloppio got his precious corpse.[22]

In England, things were more formal, though only a little. Keen dissectors would still push the boundaries at every opportunity to try and attain more, fresher and different kinds of bodies. Ultimately, this tempered the effects of the anatomical revolution; just as medical knowledge increased, so did public distrust of medical men. In 1594, Shakespeare's contemporary Thomas Nashe stirred the pot by publishing his sensational story *The Unfortunate Traveller*. In it, the hapless tourist Jack Wilton is misadventuring his way around Italy when he is captured by an unscrupulous Catholic physician who plans to use him as the centrepiece in an anatomical display (English playwrights could never resist a swift jab at Popery). Locked up and awaiting his fate, Jack muses:

> there's no such ready way to make a man a true Christian, as to persuade himself he is taken up for an anatomy. I'll depose I prayed then more than I did in seven year before. Not a drop of sweat trickled down my breast and my sides, but I dreamed it was a smooth edged razor tenderly slicing down my breast and my sides. If any knocked at [the] door, I supposed it was the beadle of Surgeons Hall come for me.[23]

Jack escapes his gruesome fate with the help of a wily courtesan, but Nashe's story captured the public imagination. After all, what was there to distinguish the vivisectionist from the surgeon, apart from good intentions? It also shows how intertwined different fields of knowledge were in this period. Medicine was part of the *ars scientia*, literally the 'art of knowledge', and sixteenth- and seventeenth-century society would have thought it strange for an educated person to be interested *only* in science and mathematics *or* art and literature. For literary authors, medicine and science were exciting tools to think with; they stuffed their poems and plays with references to astronomy, physics, anatomy and chemistry. Meanwhile, medical authors structured their works like dramas, describing the body as a mystery to be solved and new discoveries as exciting plot twists. It was a far cry from later centuries – including our own – when the structures of education and industry encouraged artists and scientists to maintain a wary distance.

Throughout the seventeenth century, the practice of anatomy flourished. There were dissections to trace the nerves and understand how living creatures experienced pain. Others attempted to fathom how women produced breastmilk or babies breathed in the womb. Famously, in the 1660s, William Harvey revolutionised understandings of circulation by showing that the heart pumped blood both through the lungs and around the body. But 'anatomy' also came to stand for a broader sort of curiosity, so that eager authors produced 'anatomies' of melancholy, plants, religion, atheism, manners, politics, war, Ireland, wittiness, writing and the art of juggling (this one so popular it went through eight editions; love of juggling apparently transcends the ages).* It was an era that defined itself by opening up, probing, experimenting and looking – endlessly, voraciously – for the truth.

---

* *Hocus Pocus Junior. The Anatomy of Legerdemain; or, The Art of Jugling*. Strictly speaking, 'juggling' in this period meant any kind of trickery or light entertainment. It took on its current, more specific meaning only in the nineteenth century.

THE
Gentlewomans Companion;
OR, A
GUIDE
TO THE
Female Sex:
CONTAINING
Directions of Behaviour, in all Places
Companies, Relations, and Conditions,
from their Childhood down to Old Age:
VIZ. As

Children to Parents.	Huswifes to the House.
Scholars to Governours.	Mistresses to Servants.
Single to Servants.	Mothers to Children.
Virgins to Suitors.	Widows to the World.
Married to Husbands.	prudent to all.

With LETTERS and DISCOURSES
upon all Occasions.
Whereunto is added, *A Guide for Cook-maids*
Dairy-maids, Chamber-maids, and all others that
go to Service.
The whole being an exact Rule for the Female Sex in General.

The THIRD EDITION.
By *HANNAH WOOLLEY*.

LONDON Printed by T. J. for *Edward Thomas*
at the *Adam and Eve* in *Little Brittain* 1682.

Frontispiece to *The gentlewomans companion, or, A guide to the female sex*. This book was published as 'by Hannah Woolley', but Woolley later denounced it as a fake. The Wellcome Collection.

CHAPTER ONE

Health Begins at Home: Domestic Healers

It is 18 October 1711. Elizabeth runs her fingers along the stacked shelves of her pantry. Here, everything is order and symmetry. She examines her medicines, noting the quantities of each in her diary:

> 114 quarts of strong cordial waters
> And of pints of cordial waters 36
> And of several sorts of syrups, 56 quarts
> Besides about four dozen of cold stilled waters
>
> In the other little closet is within the back chamber about a dozen quart bottles full, and three hampers full empty of quarts and pints; besides is one deep tub full of empty bottles, and a shallow tub full more of empty bottles.[1]

Gallons of remedies. Hundreds of ingredients. Countless hours. Each bottle had its own history – a recipe passed along from a friend or servant, potions made sweeter for her little boy or stronger for her sturdy husband. Taken together, they also tell a tale of her own long life, sixty-nine restless, discontented years. She can hardly see the labels on the bottles now. Luckily, she knows each one by heart.

***

Elizabeth Freke was a complicated woman. She was intelligent but made terrible decisions, romantic but hated her spouse. She wrote down her story hoping to pass something on for posterity, then cynically subtitled the memoirs, 'Some few remembrances of my misfortunes'. Alongside all this, she was a skilled herbalist and devoted domestic medic. Her turbulent tale gives us a glimpse into what it was like to be a woman in this period, and how female healers quietly dominated the Renaissance medical landscape.

The story began in 1671, on a 'most grievous rainy, wet day' in London's Covent Garden, where 'without the knowledge or consent of my dear father or any friend in London', Elizabeth secretly met and married her second cousin, Percy Freke, of County Cork.[2] At thirty, she was a relatively mature bride, and it was seemingly a love match, Elizabeth with her pretty round face, soft curls and quick mind, and Percy with an Irish brogue and lofty political ambitions. They'd been engaged for years, Percy carrying around the licence in his pocket.[3] But the sweetness of romance would soon turn sour. Once wed, Elizabeth realised that her husband was less wide-eyed paramour, more sharp-elbowed social climber. He started to sell off property that had been settled on Elizabeth by her devoted father Ralph, taking advantage of the laws that gave husbands almost complete possession over their wives' property. He insisted she tag along with him to his gloomy ancestral seat with her cantankerous mother-in-law or, just as bad, left her behind for months at a time while he concentrated on making his name in Irish politics. There was brief joy when Elizabeth bore a son in 1675, named Ralph for her father, but for the next two years, the infant teetered between life and death, beset by one illness after another whilst his worried mother suffered a series of miscarriages. In 1677, Elizabeth gave birth again, but from 'hard labour and several frights'; her son was 'dead borne, and lies buried at the upper end of St Giles' chancel in London'.[4] If Percy was present to share her grief, she did not think to mention it.

The Frekes' marriage was unusually well documented, but depressingly typical in other ways. Plenty of Renaissance unions were happy, equal partnerships. Oliver Cromwell's wife Elizabeth wrote to him, 'Truly my life is but half a life in your absence'.[5] Samuel Pepys, despite his roving eyes (and hands), recorded many times how he 'walked in the garden by brave moonshine with my wife' for hours.[6] However, for women, marriage meant rolling the dice on their beloved's temperament, because in the sixteenth and seventeenth centuries, they were, legally speaking, the property of their spouses. Husbands could beat their wives with relative impunity, and if she happened to die in the course of 'reasonable castigation', that was too bad.[7] Everything a woman brought to her marriage became her husband's, as did anything she managed to earn in the future. Freke should have been a wealthy woman. As the eldest of four surviving children (out of ten), she was the apple of her widowed father's eye, and he gave her numerous mortgages and deeds, as well as lump sums of cash which he urged her to keep secret. But for reasons even she could not explain, she always told Percy about the gifts, and he promptly used every penny buying up cheap land appropriated from dispossessed Catholics in Ireland. Despairing of the couple, her father finally loaned Percy the money to buy an estate of six farms in West Bilney, Norfolk. Percy got to collect the yearly rents, and in return, the manor house was 'settled' on Elizabeth, securing her a place to live.

For her part, Freke was a more difficult character than she let on in her memoirs, often embroiled in petty arguments with friends and neighbours. In 1708, in her late sixties and a widow, she was arrested. Her diary skims over the details, but we can surmise that she broke the peace in some way, maybe in a dispute over money. Writing about the incident afterwards, Elizabeth revealed her less-than-saintly side when she recorded triumphantly that God had justly punished her accuser, because 'one of [his] eyes dropped out of his head, and about three month[s] after, his wife died'.* Careful examination of her papers also reveals that she copied and revised her initial notes years later to emphasise some parts of her life story and downplay others. In her later version, Percy, once the loathed antagonist, becomes a 'dear', though errant, partner, and herself a loving and patiently devoted wife.

When marriages went awry, couples had to find what ways they could to be happy. Divorce was effectively impossible unless one was extremely wealthy, male and prepared to undergo the humiliation of telling Parliament at length about how one's wife had been unfaithful. The first divorce had been granted only a year before Percy and Elizabeth's marriage, requiring several Acts of Parliament to complete.[8] Annulments were seldom granted unless the man was shown to be impotent or one partner was already married to another when they wed. So while men diverted themselves in business, at taverns, on the hunting field or in brothels, women found means of consolation and occupation at home, refining the few spheres 'proper' to their sex into an art form.

---

* After the man's eye fell out, his son-in-law ran away to the West Indies with a mistress, leaving his daughter behind, and another man who had 'perjured' her suddenly died. Freke was delighted by the news. (Freke Papers Vol. 1, 139)

Barred from working outside the household, many middle- and upper-class women were nevertheless well-educated and literate, with a restless intelligence that needed an outlet. Medicine was that outlet for Freke; during the dark days of her marriage, she found purpose and pleasure in becoming a highly skilled and knowledgeable maker of medicines. Though her own health was poor, she delighted in nursing others, and – according to Elizabeth at least – this was how she finally reconciled with her husband. At the end of his life, Percy suffered from crippling asthma and dropsy (oedema or fluid retention), so severe that Elizabeth said eight gallons of fluid were drained from one of his legs. Something inside his body was failing. Elizabeth watched by his bedside for 'near three months', but in May 1706, she went to the West Bilney house for ten days to recover her own fading strength, thinking that 'I left him…in a fine condition and under the care of four doctors.'[9] When she returned, Percy's leg was stinking and gangrenous, and he died soon afterwards. In her view, the fault was clear: 'I told [Doctor Cousins] he had murdered my dear husband in my little absence.'[10]

***

Sixteenth- and seventeenth-century women's medical knowledge was long underestimated by historians, and many female practitioners were dismissed in their own lifetimes. Some licensed physicians looked on them as hapless amateurs, meddling in matters they could not possibly comprehend. In his 1617 text on 'several sorts of ignorant and unconsiderate practisers of physic', the Northamptonshire physician John Cotta argued that despite their 'infinitely blind good will and zeal', women would always be 'miserably perplexed' by complex cases, because they lacked a theoretical understanding of medicine.[11] His suggestion that women be banned from practising medicine was clearly designed to protect his own business from being undercut, but it was also rooted in a deeper suspicion of female power. Cotta wrote several treatises suggesting 'true and right' means of discovering witches, and he might have found older women sharing secrets between themselves a disturbing prospect, for medicine and sorcery were sometimes close kin.[12] Many witchcraft trials started with someone who had been nursing the sick finding themselves accused of causing a child or animal to become ill.* Equally, much of medieval medicine had been based on charms and rituals which trod a fine line between religion and magic, and sometimes the gleam of the arcane shone through more modern practices.

Meanwhile, even women's supporters could be more hindrance than help. In 1631, Gervase Markham, who wrote a book on household medicines, patronisingly assumed that women's medical efforts were limited to maintaining good health, not curing illnesses, and that their tinkering would be confined to their immediate family. 'We must confess', he confided, 'that the depth and secrets of this most excellent art of physic, is far beyond the capacity of the most skilful woman, as lodging only in the breast of learned professors.'[13]

Such complaints generally fell on deaf ears, and their writers dared not press the point too hard, given that some of those 'unconsiderate' female practitioners were aristocrats with more

---

* In practice (when there was a fee to be had), Cotta tended to assert that even the strangest cases of sudden illness were caused by natural rather than supernatural causes – as we will see in chapter 6.

than enough money, wit and power to bring down an insolent author. Lady Grace Mildmay, for instance, was not only a dedicated and adroit medical practitioner, but a woman accomplished in every sphere; she played the lute beautifully, could set songs in five parts and wrote a large unpublished work of 'spiritual observations' reflecting her devout Protestantism. What was more, she was such an excellent cook that when noted gourmand James VI of Scotland stayed at her home in Northamptonshire on his way to London to take up the English throne in 1603, he commended Mildmay as 'one of the most excellent confectioners in England'.[14] She could have made mincemeat of Cotta and Markham – literally.

In any case, women like Freke and Mildmay knew perfectly well that they were not simply amateurs, cunning women or 'white witches'. They were skilled, knowledgeable and sometimes ambitious, filling their medical recipe or 'receipt' books with remedies which they hoped and believed would help serious diseases like plague, epilepsy, smallpox and scurvy. There were plasters to use with a splint to set broken bones, ointments to eat away dead flesh from gangrenous wounds and lotions and potions to aid conception, prevent miscarriage, ease childbirth, help epilepsy, palsy and melancholy, heal gunpowder burns, quell cancers and even 'cure old sores on the legs, which have been of so long standing that the bones have appeared'.[15] In fact, household medicine – mostly, though not exclusively, administered by women – made up the *majority* of medical care in this period.

As ever, money played its part. This sort of medicine was often provided for free by women to their servants, family and tenants. Poorer women rendered their services for a fee, but it was still cheaper than other options. More than this, however, many people – Freke included – trusted this sort of medicine far more than that provided by university-educated physicians. Sir Ralph Verney, a baronet and prominent politician, could afford the best medical care the country had to offer, including from his physician brother-in-law William Denton. Nonetheless, in 1647, he wrote to his wife Mary asking for two prescriptions from 'Mrs Francis' and 'goodwife Greene'.[16] 'Perhaps [the] Doctor will laugh at it,' he said, 'but I know 'tis good, and have found it so myself.' On another occasion, he urged Mary not to let their ailing child be treated by physicians, but instead to contact 'midwives and old women', who, he asserted, knew more than anybody about infants and how to cure them.[17]

***

Because so many accounts of household physic have survived, we know a surprising amount about exactly how medicines were made. It was often a slow and laborious process. Cordial waters, for instance, were used for a great range of diseases, but their manufacture involved pounding, boiling and straining the ingredients, steeping them in liquid and distilling them with a limbeck or 'still'. This was a simple but effective bit of kit, consisting of a covered pot with a tube leading to a glass jar. A fire underneath the metal pot caused the liquid within to turn to steam, which travelled through the tube and condensed in the cold jar, leaving behind any impurities. Other concoctions were still more involved. One recipe for 'cock-water', a medicine designed (despite what the name might suggest) to cure consumption and fevers, asked the maker to pluck a cockerel alive 'then slit him down the back, and take out his entrails, cut him in quarters, and bruise him in a mortar, with his head, legs, heart, liver and gizzard'. The mashed cockerel was stuffed into a still with a long list of added ingredients:

> a pottle [about half a gallon] of sack [fortified wine], and a quart of milk new from a red cow, one pound of blue currants beaten, one pound of raisins in the sun stoned and beaten, two handfuls of pennyroyal, two handfuls of pimpernel, or any other cooling herb, one handful of mother-thyme, one handful of rosemary one handful of borage, one quart of red rose water, two ounces of harts-horn, two ounces of china-root sliced, two ounces of ivory shaving, four ounces the flower of French barley.*

When all fourteen ingredients were in the still, a fire was lit underneath to produce the steam which condensed in the attached glass vial. Added into this clear liquid was 'one pound of white sugar candy beaten very small, twelve pennyworth of leaf-gold, seven grains of musk, eleven grains of amber-greece [ambergris], seven grains of bezoar stone'. 'Four or five spoonfuls' of the medicine was to be taken warm every morning and evening for a month, with the author promising that 'this hath cured many when the doctors have given them over'.[18]

---

* Blue currants: probably blackcurrants, though possibly *Ribes bracteosum*, a plant from North America also known as 'stink currant'.
Pennyroyal: *Mentha pulegium*, a member of the mint family.
Pimpernel: *Anagallis*, a member of the primrose family.
Hartshorn: powdered deer antler.
China-root: a climbing plant from the genus *Smilax*, native to tropical and sub-tropical regions. It would have been available from some apothecaries.

A medicine like this took days to produce and could not be easily delegated, given that most mistresses would have been loath to entrust such a delicate process, with precious components like bezoar and ambergris, to anybody's care but their own.* Mildmay's distilling room, tucked away in a corner of the grand Apethorpe Palace, contained a huge black chest of medicinal plants and minerals, and above it, shelf upon shelf of cordials and oils, powders and pills, all neatly labelled and ready for the next patient. Some contained as many as 159 ingredients.[19] Mildmay's medicine-making went well beyond the scope of normal household duties and became nearly a full-time job, especially as she took an unusual interest in the theory behind her remedies. When a bad cold meant that she couldn't get home to prepare a certain medicine herself, she sent pages of instructions to her housekeeper, detailing how she should steep, boil and distil dozens of ingredients over the next week. 'Bess' was a dear companion to her mistress and the only person she would trust with her treasured distilling equipment.[20]

To think that every woman who practised household physic was making recipes like the one above would be akin to imagining that everyone who now possesses a kitchen is a molecular gastronomist. Our knowledge is skewed by the fact that the recipe books most likely to survive are those of the upper classes, who had leisure and money to invest in their craft. Nonetheless, this

---

* The prize for most expensive recipe surely goes to the Countess of Kent for a cure-all medicine which called for exotic components including pearls, crab's eyes and claws, white amber, hartshorn (from red deer antlers), coral, bezoar, ambergris and musk. It took a guest spot in *The Accomplish'd Lady's Delight in Preserving, Physick, Beautifying, and Cookery* (London: B. Harris, 1675), 133.

was a serious business. If John Cotta worried about domestic healers getting mixed up in witchcraft, he would have fainted at the sight of their home remedies under construction, which rivalled anything Shakespeare's weird sisters could cook up. As well as involving many kinds of excrement, and insects such as woodlice and worms, recipes regularly called for animals to be plucked, disembowelled, even cooked alive. Elizabeth Okeover's 'Oil of frogs good for all aches' required 'good store of frogs' to be put alive into an earthen pot and roasted in the oven.[21] Elizabeth Freke recommended 'the head of a coal-black cat' burnt to ashes and applied to the eyes as a cure for fading sight.[22] Numerous remedies asked for 'puppy water', an innocent-sounding substance which might mean urine from puppies, or might mean water in which puppies had been boiled.[23] Care of wounds was also an essential part of the medical repertoire, and during the bloody civil wars in the middle of the seventeenth century, Lady Anne Halkett wrote that she had treated 'three-score' wounded soldiers, even healing one whose head wound exposed the brain beneath.[24]

The ultimate test of a household medic's nerve and skill occurred on the rare occasions when they were called on to perform surgery. It is thanks to another noblewoman's diary that we get a vivid glimpse into this terrible duty. Lady Margaret Hoby was a member of the high aristocracy, moving in exalted circles. An heiress in her own right, she made three great marriages: the first to courtier Walter Devereux (brother to the queen's favourite Robert Devereux), then to Thomas Sidney (brother of the court poet Philip), and finally to Sir Thomas Posthumous Hoby (nephew of the queen's closest advisor William Cecil, son of England's ambassador to France and godson to the queen herself). Yet for all her courtly graces, she did not flinch from the grisliest aspects of medical practice. Like Anne Halkett after her, she was experienced in dressing wounds and sores; in 1600, she was tending to patients morning and evening and treating her servants who got injured on the job.[25] And in 1601, she recorded that she had performed a terrible operation on an infant:

> This day in the afternoon I had a child brought to see [me] that was born at Silpho, one Talliour['s] son, who had no fundament [anus] and had no passage for excrements but at the mouth. I was earnestly entreated to cut the place to see if any passage could be made, but although I cut deep and searched, there was none to be found.[26]

Silpho is a small and remote place even now, perched between the wild North York moors and the beaches of Scarborough. But it was only a mile or two from Lady Margaret Hoby's grand house at Hackness. Perhaps the Talliours could not find a professional surgeon – or couldn't afford one. Perhaps they simply trusted their neighbour, who, though she had no children of her own, gave medical care to many in the surrounding community, spurred on by her Protestant faith to do what she could for those who were suffering. Either way, their hope was probably in vain. 'Imperforate anus', as this condition is now called, affects around one in five thousand babies and is treatable with surgery. In this period, however, it was almost always a death sentence, and Hoby would have known as much. Later, when she scrubbed the dead child's blood from her hands, she might have consoled herself with the thought that without her he would have stood no chance at all.

***

How did household medicine become so important, and so complex? Much of what women such as Hoby, Freke, Halkett and Mildmay knew had been passed down through many generations, tracing a spidery path from mother to daughter, aunt to niece, sister to sister. In her autobiography, one of the first written by a woman, Mildmay reveals how she inherited her interest and skill in medicine from her governess, 'Mistress Hamblyn', a distant relative who had herself been brought up by Mildmay's mother, Lady Sharington. By her death at age sixty-eight, Mildmay was such an expert that she bequeathed four books full of 'recipes' to her daughter Mary, alongside over two thousand remedies on loose pieces of paper. Medical recipe books were considered precious enough to be named in women's wills, as when the Bedfordshire gentlewoman Rebecca Brandreth bequeathed books 'for surgery and physick and the other for cookery and preserves' to her daughter.[27] In fact, when Ralph Verney's mother Margaret died, she left him strict instructions to burn all her papers apart from 'medsinable and cookery books'.[28]

Texts like these are humble objects; small enough to fit in a pocket, scruffy from long use and often bound with the discarded ends of other, long-lost books. The seventeenth-century receipt book of Margaret Baker, for instance, starts in a meticulously neat hand, and gradually becomes more disorderly until the final pages have the appearance of a well-loved cookbook in the kitchen of an extremely messy chef. Happily for the historian, they are also a fossil record of triumphs and disasters, with each generation adding a new layer of experience and experiment. Ineffective remedies were tweaked or, in Baker's case, furiously crossed out. Notes squashed into the margins show how they were used; Elizabeth Okeover's book, for instance, contained an explanation for future readers that 'Some receipts in this book by mistake were writ twice over; so one of each is crossed', as well as reminders to herself to administer a vomiting medicine 'at the coming of the coldfitt'.[29] Effective recipes gained the ultimate stamp of approval, the words '*Probatum est*' – 'it is proven'.

From their tightly woven centres, these webs of expertise began to expand. The whole business of medicine was growing, and recipe collections were increasingly bolstered by input from

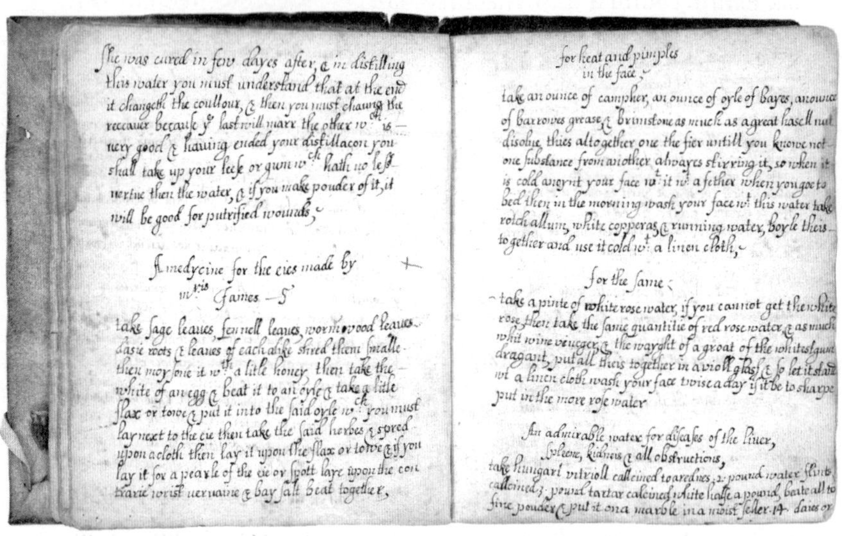

she was cured in few dayes after, & in distilling
this water you must understand that at the end
it changeth the coullour, & then you must chaung the
reccauer because ye last will marr the other wch is
very good & having ended your distillacon you
shall take up your leese or gum wch hath no less
vertue then the water, & if you make pouder of it, it
will be good for putrified wounds,

A medycine for the eies made by
mris James

take sage leaues, fennell leaues, wormwood leaues
dasie roots & leaues of each alike shred them smalle
then moysone it wth a litle honey then take the
white of an egg & beat it to an oyle & take a litle
flax or towe & put it into the said oyle wch you must
lay next to the eie then take the said herbes & spred
upon a cloth then lay it upon the flax or towe & if you
lay it for a pearle of the eie or spott laye upon the con
trarie wrist veruaine & bay salt beat together,

for heat and pimples
in the face,

take an ounce of campher, an ounce of oyle of bayes, an ounce
of barrowes grease, & brimstone as much as a great hasell nut
disolue theis altogether one the fier untill you knowe not
one substance from another alwayes stirring it, so when it
is cold anoynt your face wt it wt a fether when you goe to
bed then in the morning wash your face wt this water take
roch allum, white copperas, & running water, boyle theis
together and use it cold wt a linen cloth,

for the same;

take a pinte of white rose water, if you cannot get the white
rose then take the same quantitie of red rose water, & as much
whit wine viueger, & the wayght of a groat of the whitest gum
dragant, put all theis together in a violl glass & so let it stand
wt a linen cloth wash your face twise a day if it be to sharpe
put in the more rose water

An admirable water for diseases of the liuer,
spleene, kidneis & all obstructions,
take hungari vitrioll calcined to a rednes, 2. pound water flints
calcined 3. pound tartar calcined white halfe a pound, beate all to
fine pouder & put it on a marble in a moist seller. 14. daies or

Pages from the Receipt book of Margaret Baker, c. 1675 (f.14v-f.15r), with remedies for bad eyesight, pimples and diseases of the liver. The Folger Shakespeare Library Digital Texts.

professional midwives, apothecaries, surgeons and physicians. Mildmay's medical notebooks show she corresponded with physicians such as 'Doctor Athill', 'Dr Poe' and 'Mr Harris', who between them gave her remedies for fits following a head wound, palsy and redness in the face, as well as instructions for extracting the oil of cinnamon. With the printed word becoming ever cheaper and more accessible, knowledge-hungry women could take advantage of the glut of medical texts coming to the market. Freke hoarded books like she did medicines, collecting 'pharmacopoeias' (books of pharmacy) and more humbly titled 'herbals'. Their orderly pages, listing each plant or remedy according to its uses, were a balm to the chaos which dogged her own life. She was not alone. Books of medical recipes accounted for about twenty percent of all medical books published in the sixteenth century and the first half of the seventeenth, and became even more widespread when printing boomed after the Restoration in 1660. Authors published texts directly aimed at women, containing recipes for medicines alongside ones for cosmetics and cookery.

These books' titles promised much: 'rare and select secrets'; 'the widow's treasure'; 'the gentlewoman's companion'; 'the Queen's closet open'd'. They were aspirational as much as practical, assuming that the lady reader had an endless supply of guests to entertain and grateful acquaintances or dependants whose illnesses they could heal. But unlike some medical writing, they were not elitist. Though they might be peppered with Latin terms to keep up appearances, they weren't stuffed with arcane references or mystical speculation about the finer details of humoral mechanisms. Successful writers correctly identified that what most casual consumers – those who didn't have the skills of Grace Mildmay or the ambition of Margaret Hoby – really wanted was advice they could use every day: recipes to make ink, exterminate lice or 'get the Beetelwigges out of a man's ear' (pour into the ear water containing crushed herb-grace, have the patient lie down with their head bound in a cloth, and wait).[30] True to their roots, these texts were usually published not by licensed physicians, but by apothecaries, midwives and many of the other people physicians had tried and failed to marginalise. Of these, one woman stood out above the rest. She has been called the Renaissance Betty Crocker, Martha Stewart or Julia Child, but her name was Hannah Woolley.

The household name: Hannah Woolley

It is no coincidence that all the names we most associate with Renaissance literature and science – Shakespeare, Jonson, Newton, Dryden – belong to men. Though women wrote and passed on handwritten manuscripts, aspirations to become a published writer were viewed with suspicion and often outright hostility. To succeed required equal parts wit and determination, a path trodden by only a few single-minded individuals. Margaret Cavendish published fantastical works of fiction and serious scientific treatises, ridiculed by 'real' scientists but protected from infamy by her vast wealth and supportive husband. Bathsua Makin, reputedly the most learned woman in England, passionately argued for better education for girls. Even as the one-time tutor to King Charles I's children, she still had to conceal her gender in her published works. In the Restoration theatre, Aphra Behn turned out popular plays but was slandered and suppressed by critics who saw her as little better than a prostitute.[31] All this makes the story of Hannah Woolley extraordinary, because she succeeded not just in being a woman who wrote, but in managing to maintain a good reputation whilst being commercially successful. In fact, she was probably the first woman in Britain to support herself from writing.

Though Woolley's achievements would be groundbreaking, the beginning of her story was one she shared with thousands of other girls of the middling sort. Little is known about Woolley's life before she started writing, and the only biography provided in her own time comes from the preface to *The Gentlewoman's Companion* – a fantastically popular text, but one falsely published under her name. The story in that work presented Woolley as a child prodigy, who by age fifteen had mastered French, Italian, dancing and music, and was somehow even mistress of a small school. From there, it was claimed, she was chosen as governess to a noblewoman's family, where she also learned to make preserves and to cook, and when that lady died, she was taken into another aristocratic household, where for seven years she was 'a person of no mean authority', being 'scribe' and 'stewardess' (responsible for writing letters and accounting). Here, it was said, she also gained 'courtly phrases and graces', and acquired skill in making medicines. Later again she married and, in order to pass on her many accomplishments, once more became the mistress of her own school.[32]

Piecing together the truth from this impressive but ultimately fabricated narrative requires some detective work, using the clues scattered in Woolley's genuine works and the scant traces of her life in parish records.[33] In *A Supplement to the Queen-like Closet*, first published in 1674, Woolley gives her age as fifty-two, so assuming that she wrote the book in the year before it came out, this would put her date of birth at 1622. She was married to her first husband, Jerome Woolley, in 1646, at around twenty-four. Since she could not have been married and employed as a governess at the same time, then *if* we assume the seven-year stretch in her second employer's service is accurate, and that she spent a substantial time in the position before that (enough to gain knowledge of physic and cookery), then she must have entered service in around 1638, aged around sixteen.

Evidently, the workaday facts of Woolley's life in service had been massaged in the *Companion* to make them seem exceptional. Likewise, Woolley did run a school, but not as a teenager, and not single-handedly. When she met Jerome, he was master of the Newport Free Grammar School, eleven miles away from the grand Essex house where Hannah worked, an older, educated man with Bachelor and Master of Arts degrees from Cambridge. They struck it off as partners in life and work, with Hannah overseeing the welfare of the sixty schoolboys in their charge and Jerome teaching an intense nine-hour daily curriculum of Latin, classics, rhetoric (the art of speaking and writing eloquently) and mathematics. Following the pair through parish records, we can see that they moved from Essex to London to run another school, and Hannah had at least six children, four sons and two daughters.*

* We do not know whether all six were the children of her first husband. For a woman to have children in her early forties (when Hannah remarried) would have been unusual, but certainly not unprecedented. We also don't know how many of those children survived to adulthood. At least one, Richard, was still alive in 1674, but the rest have disappeared into unrecorded history.

Had this been the end of the story, Hannah and Jerome would long since have faded into obscurity. However, in August 1661, Jerome died. Left alone with six children to feed, the loss might have sunk a less resolute woman, but for Hannah – shrewd, imaginative and used to hard work – it was the shove she needed to begin a new phase in her life. As Jerome's health had begun to fade, she had poured her years of experience as a servant and housewife into a book called *The Ladies' Directory*, and now she took a chance, employing the printer Thomas Milbourn in the Barbican to print it at her own expense. It was a huge gamble, but it paid off. Woolley's debut was an immediate hit, and over the next fifteen years she published four more books: *The Cook's Guide* (1664); *The Queen-like Closet* (1670); *The Ladies' Delight* (1672); and *A Supplement To The Queen-like Closet* (1674). Now resident in fashionable Westminster, she would soon be widowed once again; her second husband, the gentleman Francis Challinor, died sometime before 1669. This time, however, bereavement did not bring the threat of destitution. Moreover, though being widowed twice was traumatic, it also conferred freedom, because it meant that, unlike most women, Woolley was not beholden to a male guardian. Her career was hers to direct, and her money and property hers to keep.

Woolley's books claimed to provide everything that a diligent housewife (or her diligent servants) needed to know: preserving fruit, vegetables and nuts, cooking dishes of every kind, selecting the perfect menu for any occasion, washing delicate items, making cosmetics to 'add loveliness to the face and body' and managing household staff.[34] Altogether, they were sensible, useful volumes which entered the market at just the right time. The emerging middling sort had new money, but no idea how to run the grand houses it could buy. Woolley took those aspirational women by the hand and guided them through the social maze. In all probability, not many readers of *The Queen-like Closet* actually needed to know how to make a giant artificial rock out of confectionary, or how to furnish it with snails, snakes and moss created from sugar paste, or how to make clear jellies to create the illusion of rockpools, from which a marzipan peacock could drink, or even how to arrange the whole spectacular, rosewater-spritzed edifice so that wine could pour through it and out of a little spout. The appeal of such a recipe was rather like that of TV cookery shows, designed to display the creator's skills and entertain the audience rather than providing a precise instructional guide.[35]

What really made Woolley's books stand out, though, were the number and sophistication of her medical recipes, which she presented as 'the talent God me gave', now employed 'more for my country's good, than to get fame'.[36] Woolley claimed she had begun learning how to make medicines at her mother's knee, and was now 'physician and chirurgian [surgeon] in my own house to many, and also to many of my neighbours, eight or ten miles round'. She also cautioned that she could not give details for all her medicines, because some of them were very strong and 'may as well kill, as cure', a caveat perfectly pitched to emphasise her own expertise and provoke intrigue among her readers.[37] Despite her supposed misgivings, Woolley's books kept on coming, with each new work promising scores of new and different cures. There was simply not time in one lifetime for a single person to have devised or learned all these lotions and potions, so many must have been taken from works by other, less popular authors, perhaps tweaked to make them more accessible to the everyday reader. Woolley's recipe for 'plague water' was a typical example:

> Take Rosemary, Red Balm, Burrage [Borage], Angelica, Carduus [musk thistle], Celandine, Dragon, Featherfew, Wormwood, Pennyroyal, Elecampane roots, Mugwort, Bural, Tormentil, Egrimony, Sage, Sorrel, of each of these one handful weights, weight for weight, put all these in an earthen pot, with four quarts of white wine, cover them close, and let them stand eight or nine days in a cool cellar, then distil it in a glass still.[38]

A few things stand out about this receipt. First, it was designed not to require expensive weighing equipment and to be as idiot-proof as possible. Provided all the proportions were roughly equal, the effect would be the same. Secondly, this remedy was reassuringly disgusting. Many of the herbs included were bitter, but wormwood was especially bad (as well as being poisonous in sufficient quantities).* Much like modern patients, many early modern consumers placed extra faith in medicines that were painful or tasted foul. Lastly, though it is very unlikely that this concoction could have actually cured anyone suffering from plague, it contains several ingredients that might have helped ease the symptoms of the disease in its early stages. Rosemary is now known to have antiviral and antibacterial properties, as is celandine.[39] It is unclear what plants 'dragon' or 'bural' could be here, but feverfew – misnamed 'featherfew' by Woolley – has been widely used for headaches and arthritis for hundreds of

* In Shakespeare's *Hamlet*, the eponymous hero mutters 'Wormwood, wormwood' when he views a play meant to provoke his uncle into admitting having killed his father. Not only is Hamlet's grief as bitter as wormwood, but he hopes the play will bring forth the truth just as wormwood was thought to expel worms from the guts. (*Hamlet* 3.2,181).

years.[40] Elecampane root, sorrel and borage are all traditionally thought to soothe a cough, while sage is potentially antibacterial.[41] Altogether, the recipe might have provided some relief to those lucky patients who were going to recover anyway (scientists have recently identified a genetic mutation that gives some people a forty percent better chance of surviving plague).[42] Or, it may have helped patients suffering from less dangerous feverish ailments. In either case, it would give the impression of having staved off this most terrible of illnesses.

The remedies in Woolley's works tell of a world in which serious, life-threatening diseases stalked every household, rich or poor. Each cough and cold threatened to turn into 'consumption' (a catch-all for wasting complaints such as tuberculosis). More cures were always needed for 'agues' or feverish illnesses. (At this time, malaria was still a major killer in Britain, particularly in the south-east, where flat marshlands and the filthy mud of the River Thames provided fertile breeding grounds for mosquitoes.) Painful gout struck seemingly at random, disabling the patient for days or weeks.* Then came the more prosaic ailments of everyday life: headaches, respiratory diseases, eye problems, backache and 'flux' (diarrhoea). Many remedies catered specifically to pregnant women or those trying to conceive, women in labour, or babies and infants, reflecting the dangers that this period held for both mothers and children. Even animals were not left out, with a recipe 'to cure a horse of a cold'.[43] The faith these works inspired is revealed by elegantly handwritten notes in

* Gout is mentioned frequently in medical books from this time, and may have been more common because people drank large amounts of beer and often consumed offal, both associated with a build-up of gout-causing urate in the blood.

one copy of *The Ladies' Directory*, which record the births of Mary, Williem, Christian, Elinor and a second Mary. These were all presumably the children of the unknown woman who owned the book and still trusted its recipes in 1756, ninety-four years after its publication.[44]

Woolley's blend of common sense and uncommonly astute self-promotion captured the imagination of readers in a way that few authors had managed before, let alone woman authors. Her five books went through at least twelve reprintings between them, turning a handsome profit. But where money went, trouble soon followed. For Woolley, it came in the form of *The Gentlewoman's Companion.*

Like Woolley's other books, the *Companion* offered cookery, household and medical recipes, gathered from many different sources. Much of it was Woolley's own work – she had been compiling a sequel to her popular *Ladies' Directory*, but her publisher Dorman Newman grew tired of waiting. Using Woolley's half-finished manuscript, plus new material he had gathered from other sources, he created a new and potentially lucrative text. Then, he paid a hack writer to fill in the gaps and furnish a romantic story about Woolley's life, so the patchwork *Companion* looked like a new, exciting volume in which the author supplied the secrets of both her craft and her personal history.

Where Woolley's real work was modest and private, this fake text was outspoken and opinionated. It contained a lengthy passage on the inadequacy of education for girls, arguing, 'Most in this depraved later age, think a woman educated enough if she can distinguish her husband's bed from another's.' Foolish parents, it complained, let the 'fertile' minds of their daughters 'lie fallow', while 'the barren noddles of their sons' were sent to university.[45] It is quite possible that Woolley secretly agreed with this view. Supervising hundreds of teenage boys through grammar school would have persuaded just about anybody that it was time to give female education a chance. But Woolley had spent years carefully cultivating an unthreatening, housewife-next-door reputation, and this kind of grandstanding did not fit that image. When she saw it, Woolley was furious. In her next book, she claimed that she'd told Newman she would revise the botched text for a fee. 'Truly, I dealt very friendly with him', she said, 'for I took away nothing but that which was scandalous, ridiculous, and impertinent, and put in only that which was innocent and harmless.'[46] When it came to payment, though, Newman flatly refused to pay her more than half of what they had agreed. His tactics reflect the strange state of publishing in the Renaissance, in which copyright was practically non-existent. Authors did not possess the rights to their own work once they sold them to a publisher. They might be paid for updated and expanded editions of the same text, but this was far from guaranteed. What was more, the publisher could and often did repackage parts of books for new editions, so that poems and essays would end up in different collections, or illustrations could be put to new purposes. But there was also an element here of Newman attempting to bully a female author who he probably thought wouldn't kick up too much of a fuss. Adding insult to injury, he even attached a portrait of 'Hannah Woolley' to the front of the text that was of a completely different person.* After all, surely one woman was much the same as another?

---

* The picture was probably of Mistress Sarah Gilly, a gentlewoman who died in 1659. The original engraving, by William Faithorne (after the

In this, Newman was sorely mistaken. Woolley, now in her fifties, had been orphaned, wrangled a school of sixty rowdy adolescent boys, been widowed twice and birthed six children. She was not about to be intimidated by a tradesman. While she could not stop Newman from publishing *The Gentlewoman's Companion*, or force him to pay her what he owed, she could at least set the story straight, which she did in her 1674 *Supplement to the Queen-like Closet*. The book started with a poem that did not name Newman but made it abundantly clear to Woolley's devoted readers that they were not to believe a word they'd read in *The Gentlewoman's Companion*:

> I was far distant when they printed it,
> Therefore that Book to own I think not fit.
> To boast, to brag, tell stories in my praise,
> That's not the way (I know) my Fame to raise.

Ultimately, Woolley was not only upset that Newman had compromised her carefully controlled image. She also resented the way that he'd gone about it. *The Gentlewoman's Companion* sold for two shillings and sixpence – the equivalent of a full day's work for a skilled tradesman – whereas Woolley's genuine books all sold for between one and two shillings. She had always made much of how she cured people too poor to afford a physician's fees; in the *Supplement* she told how 'I cured a bricklayer who had a sore leg by the fall of timber, and because he was poor his chirurgion

---

famous artist Peter Lely) features Gilly's coat of arms, so it was extraordinary that Newman used this picture, with the coat of arms still attached, on *The Gentlewoman's Companion*.

[surgeon] gave it over', and 'I cured a shoemaker of a sore leg, who had spent three pounds on it before he came to me'.[47] Now, even do-it-yourself instructions for her medicines were being priced out of the reach of ordinary people. Angry and disillusioned, Woolley decided to take her medical practice back to its kitchen-table roots, and invited her readers to come directly to her, to the house she shared with her son Richard on London's Old Bailey Street 'where they may have of me several remedies, for several distempers'.[48] Fittingly, given her love of hospitality, it is only a stone's throw from where the Ritz now stands.

* * *

The Supplement to the Queen-like Closet was Woolley's last book, though many works published after her death used her recipes and remedies. But for household medicine, the story was far from over. Books of household physic flourished throughout the seventeenth century and into the eighteenth, and women continued to pass along their knowledge to daughters, sisters, servants and friends. The medical marketplace was growing and changing. Physicians and surgeons would try to discourage people from treating themselves, and an ever-increasing number of practitioners offered patients a wider variety of medical options than ever before, sometimes complementing home medicine and sometimes replacing it. But as the stories of Freke, Mildmay, Woolley and others tell us, household medicine and the people who practised it were no fools, nor were they easily put off. For them, and their many patients, health always began at home.

A physician taking the pulse. Oil painting after Jan Steen. The Wellcome Collection.

CHAPTER TWO

Cornering the Market: Physicians

Bent low to see by the light of his candle, John Ward peered at the notes scrawled across the page in front of him. He really ought to be writing his sermon, but instead he was surrounded by the latest medical texts ordered from London, copying as much as he could:

> some say that contagion is a poison, some that it is a fomentation like poisons
> Contagion is a putrefaction working by way of fomentation:
> No disease void of putrefaction, though in itself never so bad, can communicate itself to others…
> It is not all putrefaction which causes contagion but there must also be a sharpness before contagion follows…
> Contagions are of 3 sorts: 1st: such as infect by touch only: 2: per fomitem [by objects]: 3: by both at a distance.[1]

There was just so much to know. Learned men like Lord William Conyers and Walter Needham, with their cool self-confidence, seemed to talk in a different language. They spoke of Hippocrates and Galen like old friends, and discussed the movements of the heavens so casually that they might be passing rainclouds. Maybe, if he could master all these arcane rules, learn the name of every medicinal plant and then somehow get to the continent to take an exam, he could join their ranks. He dipped his pen and began to write again.

* * *

John Ward was not a physician, but his diaries tell us a lot about physic in this period: how physicians trained and what they thought of other medical practitioners and their patients. He was a clergyman, caring for the souls of Stratford-on-Avon, where Shakespeare had lived fifty years before him. In fact, his sixteen surviving notebooks have probably only made it to the present day because of their brief references to Shakespeare. Famously, he wrote that 'Shakespeare, [Michael] Drayton and Ben Jhonson [Jonson] had a merry meeting and it seems drank too hard for Shakespeare died of a fever there contracted.'[2*] That line has ensured that his writings are kept safe in the Folger Shakespeare Library in Washington DC, where scholars can view them in a galleried faux-Elizabethan reading room. His scrawled, blotted handwriting gives one the thrilling sense that their author could be in the next room, having only momentarily set down his quill.

Though remembered chiefly as a vicar and Shakespeare chronicler, Ward spent far more time writing about the latest thinking in physic than about literature, divinity or anything else. He treated his own minor ailments and practised medicine

---

* Shakespeare was long dead by the time Ward was writing, but he and his wife Anne Hathaway were buried in Holy Trinity Church, where Ward was vicar.

on some of his parishioners, using what he'd learned from his extensive book collection. He liked this occupation so much that, on several occasions, he looked into becoming a licensed physician. In a diary dating from around the time that he moved to Stratford on Avon, he recorded that 'Mr Burnet has a letter out of the low-countries of the charges of a Drs. degree there which is at Leiden, about 16 pounds besides feasting the professors: At Angiers in France not above 9 and feasting not necessary neither'.[3] A couple of pages later he resolved to ask 'Mr. Yorke whether it is contrary to the laws of the land for a man in orders to get a license to practice'.[4]

Ultimately, though, his mind was too omnivorous to pursue the difficult career change. Among the pages of his medical notes were scattered all sorts of other questions, drawn from books of science and history or straight from his own over-fertile mind. Could a rabbit breed with a rat? (Yes, so it was said, and produce rabbits with long tails.)[5] Why did some men have bigger penises than others? (They grew by frequent use, and not tying the baby boy's umbilical cord too tightly would give him a head start.) What bird had the most bones? (A turkey. Turkeys also had thicker blood than other birds, were said to be extremely vicious and 'much troubled with the cramp'.) His interest roved like an errant hound, picking up one scent and then another, easily distracted by whatever new thing appeared on the horizon.

Ward's education had started him on the right path to take up medicine, if he'd wanted. In 1647, aged about eighteen, he began his studies at Oxford, where he was to stay for thirteen years, first studying for a Bachelor's degree, then a Master's, before spending the remaining years as a don (scholar). At that time, young men did not go to university to study a particular subject. The aim was to become a person of broad learning, refined eloquence and useful connections – and enjoy some good times along the way. Ward fulfilled this brief. Though he was of humbler stock than many of his fellow students, he was agreeable and ever well-meaning and had a nose for gossip, especially if it involved his social superiors. In about 1667, for instance, he recorded gleefully how 'my Lord of Exeter' (probably John Cecil, fourth Earl of Exeter) had been travelling with some friends to visit Lord Montacute when they stopped at Sir Robinson's house. They received rather too good a welcome there, for

> At Sir Robinson's house they were so whittled [drunk] that they all fell asleep in the coach: and when they came to my Lord Montacute's he came down to salute them and they were all asleep, so my Lord went up again: the footmen they stood by the coach side putting one another on to awaken them and one he durst [dared] not do it, and another durst not do it, and so the coach stood and the lords asleep in it.[6]

In 1653, Ward became a praelector, or college tutor, for Greek, and in 1654 was elected censor (another vague role involving the management of college life) for natural philosophy, which we would now call science. He began to rub shoulders with men who made their living from medicine. There was Thomas Willis, who'd practised as Charles I's physician and would go on to describe the anatomy of the brain and nerves in stunning new detail. William Conyers had gained his medical degree at the same time Ward was getting his praelectorship and stayed on in Oxford to work as a physician. One of the youngest of the group, Richard Lower, assisted Willis with his work and later made a name for himself with experiments in blood transfusion.

Ward's closest companion was Stephen Toone, an apothecary seven years his junior. The pair spent hours together in the apothecary shop, trading Oxford chatter and making medicines. In his notebooks, Ward had dozens of pages of medical recipes, and he attended sick people both as an assistant to his physician friends and sometimes in his own right. A note from about 1665 (Ward's notebooks are notoriously difficult to date and are picked up and put down over the course of about two decades) recalls that he'd examined a man called Michael Palmer who was 'troubled with a gonorrhoea and a swelling in his groin'. Ward judged that the swelling was of 'the upper part of vasa deferentia', the duct which conveys sperm from testicles to urethra, and that it might be caused either by 'putrefying' humours or by the sperm itself.[7] Elsewhere, he reminded himself 'to advise consumptive persons to feed upon ducks, and those fed with snails',[8] and made notes on how to reset a dislocated shoulder ('persuade the party to hold up her arm as high as she can, and then with a popp put it in').[9] In one particularly good month he healed Goodwife Cole when she was 'so sick that everyone thought she would have died' and received a fat pig as payment for curing Goodwife Lord's fever.[10]

John Ward's sometime practice, and his friends, provide a good example of the peculiarities of physic and physicians in the sixteenth and seventeenth centuries. By treating some patients in his capacity as a learned clergyman, Ward was operating in a similar way to the noblewomen we met in the last chapter, who provided medical care as a kind of community service. But he recognised that to scale up his practice, and to take money (instead of pigs) as payment, he would need to become qualified. This might seem obvious now, but it hadn't always been this way. In fact, the licensing of physicians was front and centre of a hard-fought battle for the soul of the medical marketplace.

***

Since the middle of the medieval period, around the thirteenth century, a handful of universities in Europe had taught medicine alongside subjects such as astrology, rhetoric and theology. Oxford and Cambridge were among those, as were Italian universities in Bologna and Salerno. At the same time, the small numbers of practitioners who treated the wealthy had begun, slowly, to organise themselves into guilds, much like the craft guilds formed by shoemakers, armourers or brewers. But, argued the physicians, their craft was more important than these others. If a shoemaker spoilt his wares, nobody died, but a bad physician was a danger to the monarch's subjects, and thus to the state itself. Ambitious physicians across the continent began to press for greater recognition, and their counterparts in England looked on with interest. Among those watching and waiting was Thomas Linacre, a scholar, cleric and physician to Henry VIII. In his portrait, Linacre wears a black furred gown and black velvet cap, slightly askew. With pale brown eyes, a thin face and his monochrome wardrobe, he's somewhat unassuming. He grasps a handful of papers in his right hand – an indicator of his learning – but more telling is his wry, closed-lipped smile, the expression of a man who knows he is the cleverest in the room. In the 1510s, Linacre petitioned Henry – then still on his first wife – to create an official College of Physicians, modelled on similar institutions in Italy. He chose his moment shrewdly. On the continent, a man called Martin Luther was challenging the papacy, and at home the troublesome but brilliant Thomas More had just published *Utopia*, his vision of a perfect kingdom. It was a time when it seemed possible that human intellect could reshape the world. Ever eager to stamp his mark on history, Henry swiftly assented to the plan. The college was founded in 1518, and in 1523 it extended its power from London to all England.

When the college gained the royal seal, its members, especially Linacre, probably thought that they would gain monopoly over medicine in England. They would take away the capacity of bishops to grant medical licences and assume total control over the accreditation of physicians. Under their watchful eye, all physicians would be university-educated scholars, able to read, speak and write in Latin and Greek, and familiar with classical works of medicine such as those by Galen and Avicenna. They would have spent a minimum of seven years, but probably longer, attending lectures and dissections and formally debating the finer points of medicine. If they had studied medicine at a university overseas, they would return to spend time at Oxford or Cambridge and have their degree 'incorporated' before they qualified for membership of the College of Physicians. It would be a golden age.

This was a lofty aim – and remarkably forward-thinking. The system has much in common with today's medical education, albeit with much more emphasis on ancient Greek and much less on practical knowledge. It was also completely unworkable in the sixteenth century. Even accounting for the fact that physicians made up a small proportion of the total number of medical practitioners in the Renaissance marketplace (something they were eager to change), there was no way that the relatively small universities of Oxford and Cambridge could produce enough of these rarefied creatures to serve the needs of a rapidly growing population. In reality, many people who practised physic held bishops' licences, for which they simply had to pass an examination by the bishop of their diocese. This allowed them to learn on the job, mixing formal education with informal apprenticeship. Others held foreign medical degrees and never bothered to obtain English accreditation – this was the option John Ward seems to have been considering when he looked into getting a quickie degree from Leiden or Angers. Of seventy-three medical practitioners operating in Norwich between 1570 and 1590, only five were university-trained physicians, with the rest a miscellany of women healers, barber-surgeons, apothecaries, bone-setters and various other specialists.[11] Throughout the seventeenth century, the number of 'licensed' physicians increased, but it was still a long way from the nationwide domination that Linacre and his colleagues had imagined.

For all the talk of standardising medicine which issued from Whitehall chambers and college lecterns, the truth was that most people simply could not afford a physician's services. Though the popular image of Renaissance England is framed by lavish banquets, starched ruffs and fine silks, the reality for all but a tiny percentage of the population was quite different. As much as half the population lived in poverty, while much of the middling sort teetered perilously close to financial ruin.[12] In the vast areas of the country dependent on agriculture, life was especially hard, as crops repeatedly failed amid the frosts of the Little Ice Age.* Sir John Wynn of Gwydirr in north Wales wrote in 1623 that he was owed £1,000 by tenants who could no longer afford to feed themselves, never mind pay rent, and 'a number die in this country for hunger, which is a lamentable thing to see…the rest have the expression of hunger in their faces.'[13] In London, in the first half of the seventeenth century, physicians charged anywhere between six and ten shillings for a visit, whereas surgeons charged around two shillings. The family would also have to find the cash to buy whatever medicines the physician recommended from the apothecary, plus more for nurses to watch the invalid and for extra nourishing food and drink. Being sick was an expensive business.

Physicians' high fees reflected their training, but they were also a marketing strategy, a deliberate attempt to position their own services as superior to those of other practitioners. It was a gamble that paid off for some – the moneyed elite were prepared to lay out eyewatering sums of money for medical care. A typical income for a physician was between £500 and £2,000, which was an extremely healthy sum, but as the physician William Denton enviously noted in the 1650s, some of his colleagues were making £4,000 to £6,000 a year. That is equivalent to around £400,000 in today's money, more than a skilled tradesman could hope to make in two lifetimes.[14] Denton even reported that the most famous physician among his contemporaries, Sir Théodore

---

* The Little Ice Age affected much of Europe and lasted nearly four hundred years, starting in the fourteenth century. In the 1600s and 1700s, the Thames repeatedly froze during winter, and the ice was thick enough that 'Frost Fairs' could be held on the river in 1608, 1621, 1677 and 1683.

Turquet de Mayerne, had left an estate of £140,000 when he died in 1655. By nimbly keeping out of politics but close to those in power, the Genevan-born Mayerne became feted across Europe, treating King Henry IV of France, James I of England and his queen Anne of Denmark, and Charles I and his queen Henrietta Maria. Oliver Cromwell *and* Charles, Prince of Wales (later Charles II) were his patients, and both paid him well for his skill and loyalty. Of course, most physicians would never in their wildest dreams have acquired such a clientele, or been rewarded so richly, but with figures like these to look up to, it was no wonder that John Ward, scraping by on his modest scholar's income, pondered changing careers.

What did doctors actually do for their hefty fees? The thing that divided physicians from other kinds of medical practitioners was their techniques for diagnosing illness. Like anyone else, physicians diagnosed patients largely from the symptoms they reported and the history surrounding the case. When the physician Edmund King visited nineteen-year-old Dorothy Battison in May 1683, he recorded in his notes that she had 'a swelling in her belly with hardness but minded it not', was pale and tired, had a weak back, pain when passing urine, and vaginal discharge. From this he concluded that she had a tumour in her womb, and prescribed her some pills which he claimed resolved the case.[15] Everard Maynwaringe, King's contemporary, paid particularly close attention to the pains his patients reported, which he acknowledged were 'the most troublesome and irksome [symptom] to bear...restless, tormenting and full of complaints'.[16] Sometimes the nature of the problem was horribly obvious, as when Gideon Harvey observed that in a patient with venereal 'pox' (probably either gonorrhoea or syphilis), 'crusty black sanious [pus-producing] devouring ulcers or sores, did eat holes into the yard'.[17] Within a few days, Harvey claimed, the entire penis would likely fall off.

In more puzzling cases, however, physicians could turn to the skills acquired during their extensive formal education. They were particularly proud of their ability to 'read' the pulse, which varied not only in speed, but in strength and 'character'. For example, the pulse of a melancholy person was said to be 'small, slow and hard'[18], while in women with menstrual problems, the pulse was irregular – Ward wrote dubiously, 'One Dr Catta once physician at Northampton affirms that Dr Langham's wife of Thornby would have an intermission of 5 or 6 pulses together.'*[19]

Pulse reading may have been unreliable, but it was far pleasanter than physicians' other diagnostic methods. Because humoralism placed so much emphasis on the body's balance, diagnosis also required examination of waste products – faeces, pus, vomit and urine. In particular, the patient's urine might contain vital clues. Was it cloudy, clear, 'thin and white', gravelly, sweet-smelling, acidic to the taste or even 'as black, as if it had been mixed with the soot of Chimneys'?[20] How did it feel against the physician's fingers, or sound when poured from one vessel into another? Thomas Willis observed that people with diabetes, which he called 'the pissing evil', had urine that was almost clear, but 'hath an honied taste' (he was correct, though he attributed it to a lack of salt rather than an excess of sugar).[21] Though some scorned this method, other physicians claimed that they could diagnose a patient using their urine

* This was probably John Cotta, the same man who wrote disparagingly about domestic healers and gave advice on how to identify witches. Make of that what you will.

A medical practitioner examining a urine flask. Oil painting after David Teniers the Younger. The Wellcome Collection.

even if they had never seen them in the flesh, and this allowed city doctors to treat wealthy clients by letter when they retreated to their country residences. The image of the learned physician peering earnestly into a vessel of urine became so associated with the trade of physic that it appeared in countless paintings, and the long-necked bottles still found in some chemist's shop windows take their origin from urine flasks.

***

Once the physician had felt the pulse, tasted the urine and pronounced his diagnosis, the treatment – such as it was – could begin. Compared to household medics, physicians were far more comfortable telling their patients what they should or shouldn't eat; presumably their high ideals and higher fees made patients more likely to heed their advice. The military medic Raymond Minderer, for instance, warned soldiers that if they insisted on risking sex with 'base women', they would be forced to undergo a cure for venereal pox, part of which was 'very lean diet, of thin broth, water-gruel, barley-broth, prunes, roasted apples, and such like, without any flesh-meat at all'.[22] When Richard Morton treated the son of his friend Reverend Steele for 'nervous consumption' – a disorder whose symptoms included thinness and depression – he counselled him to retreat to the countryside, go horse-riding and eat a special diet consisting mostly of ass's milk and soaked bread.[23] Then as now, how much notice patients took of these instructions was variable. Even John Ward, who generally idolised physicians, wrote in his diary that he'd heard of an experiment in which one group of fever patients were given a prescribed diet and the others allowed to eat whatever they wanted. Those who ate what they pleased all survived.*[24]

Dietary cures seldom did much harm, but they were tedious, and it was hard to tell if they were working. Luckily (for physicians at least), there was an alternative; the rebalancing process could be dramatically accelerated by purging the body, evacuating the bad humours in the form of blood, vomit, faeces or sweat. Purgatives were some of the most-used medicines in this period, employed by surgeons, apothecaries and domestic healers as well as physicians. They contained a wide range of ingredients from the mild to the alarming. Alexander Read's recipe, for example,

* Ward recorded a few days before this that he had given himself diarrhoea by eating too many walnuts, so he was not completely expert in dietary matters.

included senna, endives, rhubarb and rose syrup and would have produced a definite, but not dangerous, laxative effect.[25] It was used little and often to help expel melancholy humour, on the premise that whatever humour was surplus in the body was likely to come out when that body was purged. On the other hand, a singularly unappealing concoction from *Platerus Golden Practice of Physick* combined all Read's reliable stalwarts with the potent and toxic black hellebore, which the author promised could produce a 'violent' purge upwards, downwards or both at once.[26]

If self-inflicted diarrhoea and vomiting wasn't grim enough, the patient could also be bled, either by cutting a vein or attaching leeches. Bloodletting, or 'phlebotomy', was practised by many medical specialities including surgeons and some apothecaries, but physicians maintained that only they had the expertise to carry out this treatment effectively. The logic behind phlebotomy was that the blood would contain disproportionate amounts of whatever humour was causing an issue. Taking some of it away allowed the body to concoct new, more balanced blood and thus to heal itself, imitating the natural process of menstruation (or, less commonly, nosebleeds and haemorrhoids). What was more, conditions which were localised to one place, like headaches or tumours, could be relieved by letting blood in a site far away on the body, so the blood would be drawn away from the affected area. For example, young women whose menses stopped unexpectedly were considered in imminent danger of diseases caused by the humours stagnating within the body, so the veins in their ankles might be opened in order to draw the blood in the right direction. For eye problems, leeches could be applied behind the ears.[27] Exactly how much blood to take, and when, was up for debate. When Sarah Cowper found herself with a painful swollen foot and an ache in her side, she sent for 'Mr. Sloan', who let out twelve ounces (about a large mug's worth) of blood 'tho' it was night'.[28] Cowper was then in her sixties, and her doctor was probably cautious about how much blood she should lose, but John Symcotts, a Cambridgeshire physician, had no such qualms. He insisted on bleeding his seventy-seven-year-old patient forty times in less than seven weeks, claiming that otherwise 'he had been in his grave'.[29] When the German physician Christof Wirsung bled patients, he deliberately took 'as much as the patient can suffer' without starting to lose consciousness.[30]

To modern eyes, drawing blood from an already weak patient seems like madness. Some seventeenth-century physicians felt the same way. George Thomson wrote against bloodletting throughout his career, arguing that the practice was equivalent to cutting off a man's arm to remove a splinter, and was practised only by idiots and charlatans.[31] But letting blood had been a staple of medicine for hundreds, perhaps thousands of years, and patients believed strongly in its effectiveness. After all, it made perfect sense within the humoral system, and it might have provided some relief for conditions that stemmed from high blood pressure.* Some people were so convinced of bloodletting's merits that they called the physician to drain them of blood a few times a year whether they felt ill or not. Théophile Bonet, an eminent Genevan-born physician, told of one patient who 'was glad to be let blood yearly in the spring, or else she could not be well', continuing until she was

* Bloodletting is an effective treatment for certain conditions, including haemochromatosis, a genetic disorder in which the blood stores too much iron, which occurs in up to one in three hundred people of northern European descent.

eighty years old.[32] Ultimately, the practice survived into the 1800s for the simple reason that patients voted with their purses; if one physician wouldn't do the job, another certainly would.

* * *

Diet, purges and bloodletting were the bread and butter of most physicians' practices. Yet, they were only a small part of what skilled medics could offer. Physicians also prescribed and administered medicines, and they recorded their favourite remedies in instructional textbooks. Ostensibly aimed at medical students, these books were also bought by plenty of laymen and laywomen like John Ward or Grace Mildmay. Physicians piously claimed that their publications allowed people without the means to consult a doctor to benefit from their expertise, but they also made a handy supplemental income and publicised their author's most spectacular triumphs.

Among the most famed of these authors, and an early adopter of the medical textbook as marketing strategy, was Suffolk-born Phillip Barrough, whose *Method of Physick* went through seven editions in as many decades. This bumper text was among the first medical works written in English rather than Latin, and contained remedies for all ailments from the minor to the life-threatening. For instance, Barrough prescribed oil of roses and oil of chamomile to be applied for a headache arising from a hangover (along with the more useful advice of allowing the sufferer to sleep all day before taking a warm bath).[33] These were substances which most larger households would have had somewhere at the back of the store-cupboard. For migraines, however, he drew on his extensive reading of other medical authors to list ten different concoctions to be applied to the head, taken as pills, or in one case poured into the ears. They contained everything from earthworms and goat dung to 'agarick trochiscate' (a liquid made from mushrooms), frankincense and 'epithimum', a rare kind of parasitic plant also known as 'hellweed'.* Barrough's book detailed dozens of illnesses, from scalp to sole. Other ailments of the head were headaches, vertigo, memory loss, different sorts of madness, palsy, epilepsy, unconsciousness, nightmares and insomnia. Then came complaints of the eyes, ears, face and mouth and various parts of the body. The medic's travels finally ended between the toes, with carbuncles and verrucae. Barrough also devoted a whole volume of the book to the cure of venereal 'pox', or '*morbus gallicus*', showing that this disease itched at the minds of physicians as much as it did the body of patients.

In the 'letter to the reader' which traditionally appeared at the front of such books, Barrough asserted that the physician's work was God's work, wistfully imagining the great fame and respect which had been enjoyed by physicians in the ancient world. In the past, he said, people imagined men of medicine like Asclepius and Galen 'to be sent immediately from heaven... for the preservation of mankind'.[34] The reality of Renaissance physic was some way from this glorious ideal. Wealthy and educated, physicians nonetheless found themselves in a strange position. They were often the social superiors of the people they were treating, and their status as experts put them in a position of authority. Yet they were also employees, who were ultimately

---

* 'Agarick trochiscate' was likely a preparation of fly agaric (*Amanita muscaria*), a type of poisonous toadstool easily distinguished by its red top scattered with white spots.

paid by patients to do their bidding. Documents from physicians and their clients are littered with evidence of the relationship becoming strained to breaking point. Often, doctors complained that patients did not follow their advice, carrying on with the habits which had made them ill in the first place. Or, worse, they got better under the physician's care but refused to believe that their cure was anything more than chance. John Ward wrote in his book a grumble that he'd clearly heard from his physician friends:

> The 10 lepers prayed aloud in trouble, but they being once cured, 9 in ten are as mute as fishes:
>
> So is it with physicians' patients they promise fair till they are cured, but then never so much as come and thank you.[35]

For their part, patients complained when the physician's cure failed to work as they had hoped, which was often. Like Elizabeth Freke, who accused her husband's physicians of murdering him through sheer ineptitude, many doctors' best customers were also their fiercest critics. Though she consulted with a physician often, Sarah Cowper distrusted them intensely, and wrote in her diary, 'What I have lately seen by a consult of doctors confirms me in the opinion I have of their insignificancy.' She resolved, 'When necessity forces and not till then, I design to admit one and no more…Few maladies are so intolerable as these people.'[36] Cowper was constitutionally suspicious, but even when physic did seem to work, there were plenty of people willing to bet that the good outcome was due to chance rather than skill. After all, as Ward noted, 'Some says too sharply of physicians, that the sun sees their practice, and the earth hides their faults.'[37]

Conflicts between doctors and patients are well recorded because they tended to produce a lot of paperwork – letters of complaint in minor cases, and court depositions in more serious ones. What is less common, and even more precious to the historian, is correspondence between doctors and patients that shows a more typical scenario with both sides desperately struggling to cure a disease that was simply beyond the reach of Renaissance medicine. This was the tragic case of the famed philosopher Henry More's niece 'Mrs Ladde', who fell ill with breast cancer in 1674.[38]

* * *

Cancer is often thought of as a malady of the modern era, but it is an ancient disease, mentioned in Egyptian texts from 3000 BCE. Renaissance people both knew of cancer and feared its devastating effects. Its very name multiplied like tumours in the medical lexicon. 'Cancer' came from the Greek for crab, *karkinos*, 'For as a crab...whatsoever it claspeth with the claws, it holds firmly; so this grief is of a livid colour, and so girds the part which it possesses, that it seems to be nailed to the part.'[39] Elsewhere, it was called 'wolf', 'for that it is, say some, of a ravenous nature, and like that fierce creature, not satisfied but with flesh'.[40] Physicians sometimes termed it *noli me tangere*, 'touch me not', because they cautioned that cancers should not be treated, since all attempts at cure only angered the disease and made its effects even worse.

Despite this, some medical practitioners devoted their careers to the pursuit of a remedy for this most terrible of diseases. Each had their own, fiercely held, beliefs, but they can be roughly divided into four schools of thought.

The first school attempted to correct the humoral imbalance which had brought about the disease. Henry More, well versed in humoralism, held that the most common cause of cancer was an excess of melancholy humours. What More did not mention (or did not know) was that particularly aggressive cancers were thought to stem from 'adust' melancholy, a noxious substance created when regular melancholy was 'burnt' either by becoming somehow stopped up in the body (for instance, by a bruise to the affected part) or by meeting and mixing with boiling choleric humours. Harking back to the dietary recommendations that underpinned so much of Renaissance healthcare, conservative treatment of cancers might therefore include a light, 'cool' diet, rich in green vegetables and fish and sparing with red meat and salty foods. Most physicians included this kind of advice alongside whatever else they suggested.

The second route favoured medicines containing ingredients with an 'affinity' to the disease. To early modern people – as to many modern patients – cancer felt like a *creature* attacking the body, a parasite. Perhaps, then, it could be cured with medicines which contained bits of the disease's namesakes: crushed crab, woodlice, worms or, in one case, a wolf's tongue.[41] Likewise, domestic healers would take 'the flesh of an hen, chicken, pigeon, whelp [puppy], or kitling [kitten], cut asunder', and apply the warm carcass to the ulcers caused by cancerous tumours.[42] The cancer would supposedly pause devouring the body of the patient in order to 'eat' the more tempting meat, offering them a temporary reprieve from its quasi-sentient attack.*

The third option was application of very strong 'plaisters' (thick ointments) containing arsenic or mercury, which sought to kill the cancer by corroding the flesh in which it was contained. These kinds of treatments were known to be dangerous, and were rejected by some physicians for being only 'able to increase the pain, fever and all the symptoms, to the dejecting of the powers, the wasting and consuming of the body, and the hastening of death'.[43] However, some doctors, and patients, saw these substances as a necessary evil, a sort of early chemotherapy which would hopefully kill the cancer before it killed the sufferer. Only such potent remedies possessed 'force to consume the evil humour'.[44] This is the route Mrs Ladde ultimately took.

Finally, there was the terrible and radical option of surgery. While this was the only treatment which could feasibly destroy the cancer, it was as likely to kill as cure.**

In 1674, when his niece fell ill, More was sixty years old, with no wife or children of his own. To the intelligentsia, he was known as a gifted philosopher, churning out works of theology and metaphysics. To his friends, he was kind, humble and humane. It was said that during his time at Christ's College,

---

* One well-reputed physician, Daniel Turner, claimed to have heard a story wherein 'when [doctors] held a piece of raw flesh at a distance from the sore, the wolf peeps out, discovering his head, and gaping to receive it'. Turner dismissed the tale as 'villainous', but it shows how ideas about disease sometimes slipped between metaphorical and literal. (Turner, *De Morbis Cutanels* (1714), 76.)

** The details of the mastectomy operation, and the courageous women who underwent it, appear later in this book.

Cambridge, More had been known as 'The Angel of Christ's', and in 1666, he gave up his comfortable living at the benefice of Ingoldsby, near Grantham, to another churchman who faced destitution. Among his closest friends was Lady Anne Conway; his sometime pupil, now an equal. They were avid correspondents, and through their letters we can trace the hope and tragedy that awaited his kinswoman.

In July 1674, More wrote to Conway asking a favour. He wanted to consult her physician friend, Francis van Helmont, on 'what is most proper and effectual for a cancer in the breast, near ulceration'.[45] He would not trouble the doctor, he explained, were it not that the patient was a relative, and he felt duty-bound to procure the best advice money could buy. Francis was the son of the famous chemist Jean Baptiste van Helmont, who established the radical medical theory of 'Helmontianism'. While lacking some of his father's creative intellect, he was a renowned (sometimes infamous) alchemist and physician in Europe, roving the continent peddling unorthodox religious and scientific ideas which landed him in the hands of the Inquisition for eighteen months in 1661–2 (ever alert to medical gossip, Ward recorded in his diary with some surprise that the Inquisition had not managed to kill Helmont with their tortures).[46] In 1671, he settled down for a time at Ragley Hall, Lady Conway's grand and beautiful Warwickshire residence.*

* There, between doing diplomatic work for Princess Elisabeth of Bohemia and rubbing shoulders with the great and good of the day, he served as physician for Conway. She was an intellectual powerhouse, as learned on the subject of Cartesian philosophy as any man, but suffered with debilitating migraines which the country's best doctors – Francis included – failed to cure.

The 'Helmontian' school prided itself on dismissing old-fashioned Galenic ideas about balancing the humours, promoting instead strong chemical agents to support the body's 'vital principle', which was something like an amalgam of the immune system and the soul. Their ideas ranged from the forward-thinking to the bizarre. But by the time Mrs Ladde fell ill in 1674, their less eccentric notions had been incorporated into the regular practice of many physicians, making a hybrid medical model from which doctors took whatever seemed to work. Helmont sent a box of medicines to More, which he sent on to his niece's local physician (and another of More's ex-pupils), Dr Clark. Mrs Ladde lived in Grantham, far from the metropolis, and More hoped to impress her with Helmont's credentials. He told Lady Conway, 'I have writ to Dr. Clark to possess my kinswoman with as high an opinion of the medicine as may be, and have writ to her myself to certify her of the [reputation] and fame Monsieur Vanhelmont has over all Europe, if it may contribute to the efficacy of the medicine.' Forebodingly, he added, 'I pray God give good success to the takers of them'.[47]

More did not record exactly what Helmont prescribed. But it must have been potent, because within a week of taking the medicine, his niece 'Mrs Ladde' was 'ill at her stomach, hot and thirsty, with frequent shootings in her breast, and not only on the cancered part, but likewise round about it'. She also had 'many little angry pustulats [pustules], first red, afterwards maturated on their heads'. Both symptoms are consistent with what medical texts of the time describe as the side-effects of arsenic or mercury treatment, and both More and his niece were clearly rattled: he wrote to Conway urging her to ask Helmont for further advice 'with what convenient speed may be'.[48]

Somehow, the doctors persuaded Mrs Ladde to continue her 'cure'. By December, however, there was another, more serious problem. Though Dr Clark had said nothing, More had heard through his nephew that she was gravely ill. She'd been advised to keep going with the medicine, he wrote, but she was wary, because it had 'much increased the bigness of the sore which is all in such fretting distemper'. Faced with this report, More began to suspect that Dr Clark had concealed the true gravity of the situation: 'I believe he does not deal so openly with me as he should, out of a niceness to displease me, Because of my great opinion I have expressed of Monsieur Vanhelmont.' In all, he told Lady Conway, 'I am something at a loss what to do in the case and dare [not] over sensibly press the use of the plaster upon my niece. For fear of the worst.'[49]

More was torn between his faith in Helmont's theories and the mounting evidence that Mrs Ladde was getting worse. As he prevaricated over what to do next, however, the situation deteriorated, and at the end of the month, he wrote to Lady Conway again, this time transcribing directly out of a letter received from Dr Clark:

> I looked on Mrs Ladde's breast yesterday, and I think the tumour is not much altered, but the ulcered part is rather broader though no deeper. She keeps the medicine to it and can better undergo the pain thereof than before… And indeed considering the nature of that distemper I wonder it is no worse…which makes me hope that the medicine is proper for it.[50]

Optimistic, foolish or simply out of his depth, Dr Clark tried to convince More that the patient's rapidly growing open sore was not a big problem. However, a month later, Mrs Ladde was suffering worrying symptoms once again; after waking up in the night, she 'was taken with a violent pain in her head with almost a blindness and stupidity'. She cried what she estimated was a 'porringer' [porridge bowl] worth of tears, which Dr Clark thought had helped fend off an apoplexy. More could do nothing but wait and 'hope the danger is over, I hearing nothing from them to the contrary'.[51]

By this time, Helmont evidently saw that things were beginning to go badly awry, and had started to distance himself from the case, telling Clark to judge what he thought best on the spot. No more letters about More's niece from the next eighteen months survive, but by the spring of 1676, he told Lady Conway that he was travelling to see Mrs Ladde, as she lay dying.[52]

More's letters show how complicated the relationship between a patient and their doctor could be. His niece trusted his judgement, and he trusted Helmont. Even when the medicine seemed to be doing harm, he wrote that 'what ever it be I shall account myself much obliged to Monsieur Vanhelmont for his good will'.[53] But More's niece discontinued her treatment when it became clear that it was making her symptoms worse, and she, her brother and More all evidently lost faith in Dr Clark, who was less esteemed. The power dynamic between More, his kinswomen and the two doctors was a complicated dance in which each one tried to negotiate their financial and humanitarian obligations to each other. Medicine never operated in a vacuum, but depended on the social world where it was embedded, complete with love, lies and petty rivalries. Nowhere was this more powerfully illustrated than in the life of one argumentative, ambitious physician, George Thomson.

The plague doctor: George Thomson

In 1665, the worst attack of plague in living memory swept through Britain, devastating its population. It started slowly, creeping in with the sticky summer heat. In the second week of May, nine people from four parishes across London and its outskirts had died. In the second week of June, the death toll had crept to forty-three, with 7 of 130 parishes reporting cases. But by sweltering midsummer, nowhere was safe. In the week of 19–26 September, 7,165 people died in London alone – 1.5% of the city's entire population perishing in just seven days.[54]

It was not only the capital that suffered, for there were tragedies in every corner of mainland Britain. Famously, the occupants of Eyam, in the Peak District, made the altruistic decision to quarantine their village when the plague arrived, rather than risk infecting their neighbours. Of the 800 inhabitants, 260 died over a fourteen-month period, but they succeeded in stopping the disease – which had likely arrived via lice in a parcel of wool – from spreading to the larger area. However, London was by far the worst affected region. Much of the city was cramped and filthy, 'environed with ditches [and] stinking gutters', with livestock in the streets and raw sewage emptied directly into the rivers.[55] Most food and other goods came in by carts, which crowded together at the gates to the city walls and ground to a halt on London Bridge, the capital's main thoroughfare, which was only twenty feet wide. At the busy docks, ships from all over the world brought exotic goods, but also vermin and sickly sailors, the latter seeking out a warm body and a stiff drink in the city's bawdy houses and taverns. Altogether, it was the perfect environment for the fleas and body lice which carried the plague pathogen, *Yersinia pestis*, to roam freely from host to host.

Plague comes in several forms, including pneumonic and septicaemic, and it is likely that a deadly mixture was at work in 1665.* By far the most common, however, was bubonic plague, which caused fever, terrible headaches and the distinctive enlarged lymph nodes, or 'buboes', which gave the disease its name. Victims shivered and sweated, then vomited blood while their skin died on their bodies, turning black and gangrenous. Once infected, they were unlikely to survive more than a week. Britons were well aware of this grim statistic – plague had struck several times in the preceding decades, though on a slightly less devastating scale. The Bills of Mortality, which began in 1603, recorded that 29,402 people had died of plague in London that year.[56] In 1625 the number was 41,313, and there was another smaller outbreak in 1635.[57] So it was that when this plague started to spread faster and further than previous epidemics, anyone who could afford to leave London fled for their lives, abandoning the city to the poor, the dying and the dead. As cases began to soar in early June 1665, Samuel Pepys watched in horror, writing in his diary:

> This day…I did in Drury-lane see two or three houses marked with a red cross upon the doors, and 'Lord have mercy upon us' writ there – which was a sad sight to

* If left untreated, bubonic plague has a mortality rate of between forty and seventy percent. Septicaemic plague spreads through the bloodstream, while pneumonic plague affects the lungs. The mortality rate for these varieties is nearly one hundred percent.

> me, being the first of that kind that to my remembrance I ever saw.[58]

By July, people were so afraid of anyone or anything coming from London that when Pepys went out of town to try and arrange a marriage, he had to lie to people about having been in the city. Over the summer, he decamped from his house in Seething Lane in central London to Greenwich, then a separate small town, and his wife Elizabeth was installed a few miles further away in Woolwich.* In August, he was in London and nearly tripped over the corpse of a plague victim left in the street. 'How few people I see', he remarked, 'and those walking like people that have taken leave of the world.'[59]

Among those who escaped London were the city's physicians. With wealthy clients before them and a seemingly unstoppable disease at their backs, it seemed only logical to make for the relative safety of the countryside, though the mass exodus damaged their reputation for years to come. A treatise released by the College of Physicians in 1665 optimistically posited that a mere 'six or four' doctors needed to remain in the capital, each with two apothecaries and three surgeons to assist them.[60] This proposal, along with the sound but impractical advice that all victims be buried far from the city, suggests those at the top of the college tree failed to understand just how many people were stuck in the capital, or how many were dying each day. Yet a few doctors remained. Of

---

* This circumstance meant that Pepys did not have an entirely bad time during the plague outbreak. Socially distanced from his wife's watchful eye, he was free to indulge in as much drinking, socialising and womanising as he pleased, and at the end of 1665, he wrote, 'I have never lived so merrily…as I have done this plague-time'. (Pepys, 31 December 1665)

these, the most well-known was the maverick George Thomson: ex-soldier, ex-prisoner, devotee of chemical medicine and thorn in the side of the medical establishment.

Thomson could not have been further from the genteel, Oxbridge-educated men with whom Thomas Linacre had once hoped to populate the College of Physicians. As a teenager, he'd learned much from the apothecary Job Weale, a rebellious figure who was fined huge amounts by the college for refusing to obey their strictures. Thomson then travelled in France, fought in the Royalist army in the civil wars, and was captured and briefly imprisoned before finally gaining his long-delayed medical qualification – not at the hallowed halls of Oxford or Cambridge, but at the relatively new University of Edinburgh, topped up with medical study at the University of Leiden (the same place John Ward had thought of going to gain his MD). Setting up in Romford, Essex, he attempted, unsuccessfully, to make his name in the profession by removing the spleen of a dog (it lived for two years and three months afterwards, proving Thomson's theory that the organ was unnecessary). By the time he finally landed in London in 1659, Thomson was forty, unmarried, low on funds but armed with life experience and remarkable grift. A less confident man might have kept his head below the parapet, especially since, despite his training, Thomson was not licensed by the college to practise medicine in London. But Thomson was already bruised by the lack of recognition he'd received for his dog splenectomy and was not about to hide his light under a bushel. Soon, he published his first work, not a book of remedies or a new anatomical study, but a scathing attack on traditional Galenic humoralism. Thomson was a zealous devotee of the new Helmontian school. He rejected bloodletting as superstitious, dangerous nonsense, by which 'the patient is sent piece-meal to his grave', and the men who practised it – the cream of the medical establishment – as 'rotten'.[61]

When the plague broke out in earnest in 1665, Thomson found himself in the heart of the storm. His house in Duke's Place 'nigh Aldgate' was five minutes' walk from one of the city's mass graves at the church of St Botolph without Aldgate. He would have seen and smelled the carts rolling past laden with bodies and heard the church bells tolling for the dead – at least, until the bellringers died and the city fell silent. The grave in St Botolph's churchyard contained over a thousand bodies, as Daniel Defoe described some seventy years later:

> A terrible pit it was, and I could not resist my curiosity to go and see it. As near as I may judge, it was about forty feet in length, and about fifteen or sixteen feet broad, and at the time I first looked at it, about nine feet deep; but it was said they dug it near twenty feet deep afterwards in one part of it, till they could go no deeper for the water; for they had, it seems, dug several large pits before this.[62]

Plague pits served a purpose; they got the bodies out of the way and avoided further outbreaks of disease. But they also piled more horror upon an already traumatised city. Most people believed fervently that they would be resurrected in the same body at the Day of Judgement, and mass graves raised the disturbing possibility that your risen body might get mixed up with someone else's.[63] To Thomson, they were both a spur to action and an opportunity to nail his colours to the mast. In 1666, he published *Loimotomia, or, The Pest Anatomized.* In it, he claimed that he'd been 'wounded by the venomous arrow of the direful pest' three times. On the first occasion he'd been treated 'to my sorrow' by a Galenic physician, and nearly died as a result.[64] The next two times, he used his own chemical medicines, and the arrows of disease were quickly extracted. He now felt that it was his duty – and the neglected duty of his brother physicians – to do what he could to save his fellow Londoners. Emboldened by his beliefs and energised by his clashes with medical authority, Thomson moved freely among the sick, 'conversing with my patients, and giving them effectual remedies, prepared with my own fingers, opening their buboes… by the operation of my own hand'.[65] As far as any sane person could be unafraid of the plague, he was unafraid.

Contemporary thinking on the causes of the pestilence was, to put it charitably, confused. In a general way, people agreed that the 'visitation' was a punishment sent by God in retribution for humanity's wickedness, and at least as much literature was devoted to religious speculation on the subject as to medical observations. There was extra reason to believe that the disease was supernatural in origin; in late 1664, Pepys recorded that London's coffee-houses were abuzz with talk of a bright comet which had appeared in the night sky and would continue to be observed for many weeks, with another in 1665. Such phenomena were widely agreed to be portents of disasters to come. When the extent of the unfolding epidemic became apparent in the summer of 1665, King Charles II sent forth a proclamation ordering that the first Wednesday of every month should be a day of 'fasting and humiliation', hoping to appease a wrathful Creator.[66] Alongside these heavenly causes were more material ones. Many blamed the advent of plague on 'corrupted air', especially since that summer was exceptionally hot. Corruption came from astrological events such as comets, from the bowels of the earth via volcanoes and geysers, from stagnant water and from rank unburied carcasses. Physicians also recognised the role of person-to-person contact in the spread of the disease, though exactly how this happened was somewhat mysterious. William Kemp speculated that sick bodies sent out 'fumes and steams'.[67] Gideon Harvey advised that the 'miasma' of the disease could enter the body either through the breath or via the pores.[68]

Though these assessments entirely missed the role of fleas and lice in spreading plague, they still produced some sound advice in the form of a Renaissance lockdown. 'Shun all publick meetings', urged Harvey, and 'avoid passing close, dirty, stinking and infected places' such as alleyways, especially those where household bedding was hung out. If travelling and staying in an inn, lie on the floor rather than on the mattress; if in a boat, spurn the straw seating provided for comfort.[69] All these guidelines effectively protected the reader from the most likely places they could pick up infected fleas from rats or lice from other people and their possessions. Likewise, the government ordered infected households to be shut up for forty days and prohibited trade between infected towns. 'Searchers' employed by the state inspected dead bodies for signs of the plague and committed the infected to the dreaded 'pesthouses', hospitals which locked the inhabitants away from healthy citizens. In theory, it was a life-saving set of rules, but it was horrific for those shut in infected houses, waiting for the sickness to overtake them. Local authorities had to contend with quarantined people breaking or bending the rules, and the physician William Kemp complained, 'I have observed that most of these people are extreme ignorant…Or else they have no good nature, or kindness for mankind; or else they are exceeding covetous; or such as care not much to be rid of some of their relations'.[70]

Thomson's understanding of plague was similar to that of more traditional physicians. It was, he argued, spread through a sort of 'venemous Gas', 'which though small in quantity, yet operates stupendiously' (he was very fond of long words, even if he had to invent them).[71] But he differed from Galenic physicians in believing that the body's workings were controlled by a mysterious quasi-sentient life force called the 'archeus'. The symptoms of plague were caused by the *reaction* of the body's archeus to the 'atoms' of disease.* These 'poisonous atoms that lurk in the pores of the air…entering into our bodies unfelt, unseen, cannot of themselves make the pest, unless our archeus become an efficient and formal cause of its quiddity [essence]; so that it hath its immediate being from our vital spirit'.[72] It was, inadvertently, a fair description of the immune system. However, as so often with medical texts, it was unclear exactly what this abstract line of thinking actually meant for people suffering from plague. What exactly was the archeus doing to turn the 'poisonous atoms' into deadly sickness? And why was that sickness so intractable?

Casting around for answers, Thomson decided to probe further into the matter by taking the extraordinary decision to conduct a postmortem on a teenage plague victim, probably the servant of one of his many acquaintances. The scene he described was hellish. The youth's body lay in the dim, close room in the house where he had died, deathly pale but covered in black swellings.

* Thomson did not know of atoms in the modern sense, but used the term to describe very small particles of matter.

Physician George Thomson dissecting a plague victim. Line engraving, 1666. The Wellcome Collection.

Thomson set fire to 'brimstone' – that is, sulphur – in order to try and protect himself from the body's pestilent vapours, creating a foul-smelling gas that irritated his eyes and throat. The corpse's interior was even more foul; chyle in the stomach of the dead youth was black, and his blood seemed to have coagulated in his veins. To Thomson, this was fuel for his campaign against blood-letting, for he believed that his findings showed how the blood was responsible for making digestive juices in the stomach. He also noted that the body was still warm twelve hours after death, which he put down to the extreme agitation of the boy's archeus – though as usual, exactly what 'agitation' meant was never made entirely clear by Thomson or other Helmontian physicians.[73]

However, the most important findings did not come from the dissection itself, but its unintended consequences. Shortly after completing the postmortem, Thomson came down with the plague yet again. He was surprised by this turn of events (probably the only one surprised), for there had not been much smell from the corpse, and, like other physicians, he believed that it was the 'emanations' that were dangerous. In fact, he complained, he'd experienced many 'abominable, loathsome, noisome smells, from sores, carbuncles, the tainting respiration, and fetid expiration of numerous living bodies' without getting infected before.[74] Having no idea that the bacteria could have lodged under his fingernails, or the infected fleas might have jumped from their expired host to a new and tastier one, he concluded that the problem was that he'd had his hands in the young man's gore for so long that the 'venom' had seeped in through his pores. In any case, it was an ideal opportunity for Thomson to test out his chemical medicines on himself, and on the other members of the household who were now also in peril.

Thomson's regimen started off well, with common sense recommendations: eat and drink well, exercise, sleep much and try to retain an 'undaunted firmness of mind'.[75] While his belief that burning sulphur could 'exhilarate and dilate the vital spirits' was wrong, it was almost certainly justified by experience, since sulphur dioxide can be used to kill fleas.[76] From there, however, his treatment plan rapidly turned arcane. The Helmontian school believed in the power of chemical medicines to destroy disease 'atoms' and support the beleaguered archeus. When the pest had taken hold, possible remedies included 'the saline spirit of blood, bones and urine', 'salt of tartar' and 'oil of amber', as well as the old favourite, mercury.[77] Thomson claimed that in his previous illness, he had found relief from the dissolved flesh of snakes, though it was unclear if he used that now.[78] Strange even by Renaissance standards, these ingredients reflected the desperate race to find something that would stop this disease. Virtually everything that could be powdered, eaten, drunk or smeared on the body appeared in a plague remedy somewhere, from walnuts to wormwood, opium to unicorn-horn. Mercifully, the official advice from the College of Physicians was for once to avoid using the 'three great remedies' of bleeding, purging or inducing vomiting.[79]

Thomson's suggestions for a cure were no better or worse than those of his Galenic contemporaries, but they were certainly novel. He had heard how 'Some things outwardly applied and worn are much commended, as the emerald and sapphire'. When drawn in a circle around the plague buboes, these allegedly 'do magnetically extract the virulence and malignity in the body'.[80] Thomson was not rich enough to have loose jewels lying around, but he came up with an unlikely alternative. In

Loimotomia, he advised that a dead toad, or 'Bufo', could be ground into powder along with its vomit and made into tablets, which when applied to the body of the plague sufferer helped to extract the pestilent matter in much the same way as a precious stone. This was one of many strange beliefs attached to toads; others being that they had a stone in their head that was an antidote to any poison, and that toads and spiders were mortal enemies. On this occasion, however, Thomson explained that the reason behind the toad's efficacy was its hatred and terror of humans, so great that if a person stared at a toad very hard for fifteen minutes or so, 'it will shortly die with very terror' (sadly, he did not say how he had found this out).[81] This terror, remaining as if it were fossilised in the toad even after death, somehow worked to overtake the fright that the archeus felt about the plague infection…though quite how this worked was never very clear.

Toad or no toad, all the members of the household recovered from their illness – a remarkable feat. Having had the plague several times, Thomson must have had some natural immunity, while the survival of the whole family suggests that they may have encountered a milder strain. Thomson survived to continue baiting the College of Physicians for another twelve years.

Thomson's ideas seemed almost as strange to his contemporaries as they do to modern readers. But while the wider uptake of 'chemical medicine' was never very enthusiastic, it contributed to later medical developments in ways that neither its detractors nor its champions could have foreseen. True, the Helmontians' faith in the power of objects like jewels and dead toads harked back to the folkloric, magical medicine of earlier centuries. It was no coincidence that they sometimes found themselves on the wrong side of the religious authorities. However, in theorising that illness arose from disease 'atoms' rather than imbalances in the humours, Thomson and his fellow chemical physicians were a little closer to the modern model in which bacteria and viruses cause many diseases, an idea which would not emerge for several hundred years.

Most importantly, rebel physicians such as Thomson helped to keep the College of Physicians on their toes. By the time the plague finally started to subside in 1666, Thomson had lost several of his friends and colleagues, all middle-aged men who shared his devotion to the new science: Joseph Dey, Thomas O'Dowde, Robert Turner and his beloved mentor, George Starkey. Starkey was another believer in the power of toad-based medicine, but when he realised he had contracted the plague, he'd just drunk a large quantity of beer. Apparently, this was his undoing; the beer caused his pores to close so that the toad remedy could not enter his body to do its work, and he died the same night. Nonetheless, there were still enough Helmontians left in London to prove a real challenge to the Galenists, and as the pestilence died down, they redoubled their efforts to form a Society of Chemical Physicians which would rival the College of Physicians. For a while, it looked as if they might succeed, and Thomson put the two groups' response to the plague epidemic at the centre of the debate. In the army, he pointed out, 'deserters and runaways' were punished with death, but the way that the collegiate physicians had abandoned London to its enemy the plague was just as bad.[82] In fact, he argued, 'It were not amiss for the magistrate to force these fugitives to return to do their duty, and compel them to visit the sick, and to take (as well for cure as prevention) those unnecessary antidotes they have left behind according to their wise directions.' If college doctors refused to trust in their own medicines, then 'I hope the sage magistrate will be pleased hereafter to favour signally the true artists, and to reject these counterfeit and useless doctors, making them (as they really deserve)...a laughing-stock.'[83]

In the end, Thomson's dream of a Society of Chemical Physicians would never become a reality. Those who'd died during the pestilence had been some of the movement's leading lights, and their loss left the remaining men divided, squabbling over details. But there was some progress. In 1667, a year after the epidemic had passed, Charles created a 'chemist-in-ordinary' post at his court, followed a few years later by that of 'chemical physician to the king'.[84] Neither job went to Thomson. Permanently angry and constitutionally unable to bite his tongue, he would have been an ill fit in Charles II's merry, libertine court, and instead he put the profits from his writings toward a new laboratory in London's smart Soper Lane, right on the route of the Lord Mayor's annual procession. Thomson's career illustrates the complexity of the medical marketplace. His enemies considered him a lunatic, but he considered himself to be the future of medical orthodoxy, and his partial success was also a win for other specialities such as apothecaries and surgeons, who successfully eluded the College of Physicians' attempts to bring them into line.

The 1665–6 epidemic would be the last great outbreak of plague in Britain, and it left behind a traumatised nation. Almost everyone had lost neighbours, friends, relatives. It did not make people give up on medicine, but it made them more determined than ever to choose their own path through the ever-growing medical marketplace. Thomas Linacre had dreamed of a black-and-white medical world, physicians versus quacks. But as medical ideas and practitioners proliferated, the reality was something far more subtle and strange.

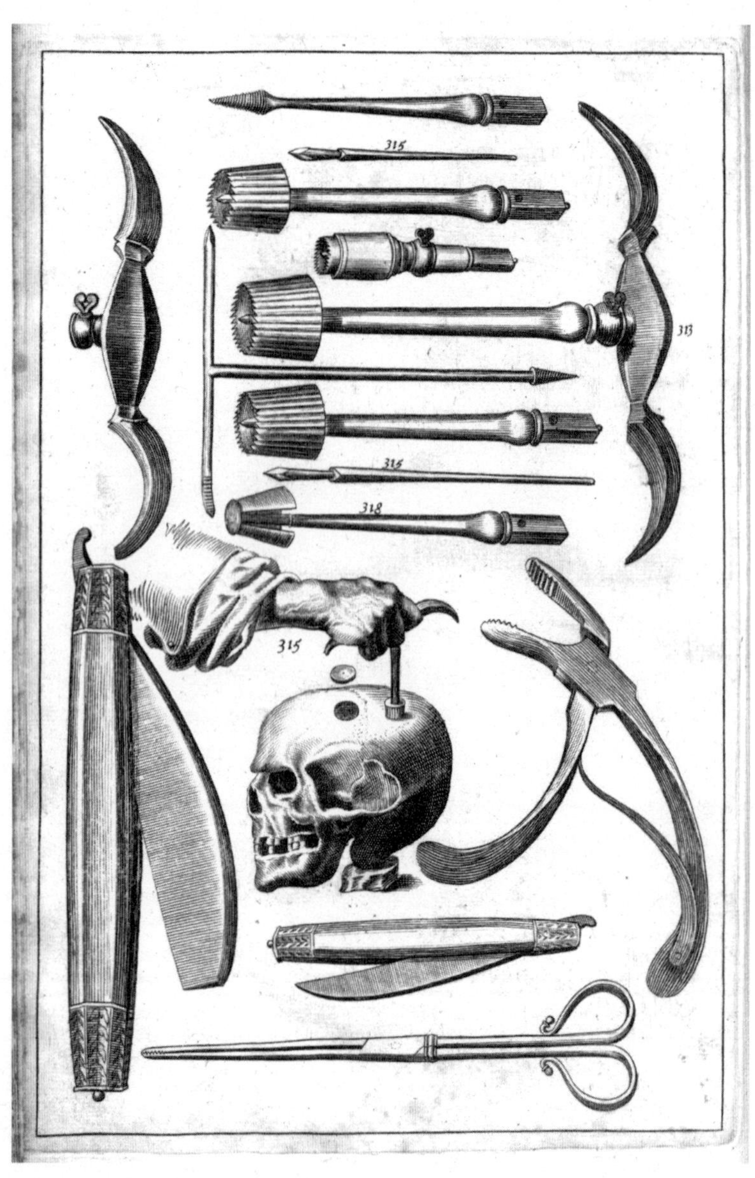

Instruments for skull operations, from *The surgeons mate or military and domestique surgery* by John Woodall. The Wellcome Collection.

CHAPTER THREE

Right Place, Right Time: Surgeons

It was a good day for John Ward, and a very, very bad day for Mrs Townsend.

As she entered the room, she was greeted by three faces, pale with worry. Mr Clark had come from Bridgnorth and Mr Linch from Sturbridge, both towns two days' ride from her home in Alveston, Stratford-upon-Avon. There were other surgeons nearer, but not everybody wanted to get involved with an operation where the odds seemed so stacked against them.

The third face was that of Reverend Ward, Mrs Townsend's vicar and amateur medical enthusiast. Townsend knew he was there to ogle at the surgery more than to offer her spiritual support, but in these circumstances, it didn't hurt to have a man of the cloth on hand. She lay down, said a prayer, and the operation began. As Ward recorded,

> First they cut the skin crosse and laid it back, then they worked their hands in, the one above and the other below and so till their hands met, and so brought [the tumour] out: They had their needles and waxed thread ready but never used them, and also their cauterizing irons but they used them not: she lost not above [6 ounces] of blood in all.[1]

Finally, the bulk of the tumour was out, but this was not enough. If anything remained, the cancer would come back and her suffering would have been for nothing. The next day Clark, Linch and Ward returned, and they brought along a new surgeon, Dr Needham. Clark and Linch reopened the wound and sliced away more flesh. The next day, the same; 'Every time they dressed it they cut off something of the Cancer that was left behind; the chirurgions [surgeons] were for applying a caustic, but Doctor Needham said no not till the last, since she could endure the knife.'

It was more than anyone could bear, yet somehow, Mrs Townsend did. Ward wrote in his diary, 'One of the chirurgions told her afterwards that she had endured so much that he would have lost his life ere he would have suffered the like: and the doctor said he had read that women would endure more than men, but did not believe it 'til then.' She'd surprised the surgeons, the vicar, and even herself. Though Mrs Townsend had far to go, she had survived the worst of her ordeal.

* * *

A mastectomy is a daunting prospect today, despite all our medical advances. It is almost unthinkable that in the seventeenth century a person would volunteer to go under the knife without anaesthetic, not just once, but over several days, each time risking their life and enduring incredible pain. Yet this is exactly what Mrs Townsend of Alveston did, and her operation was documented by her local vicar, John Ward, the would-be physician we met in the last chapter.

Mastectomies were unusual in two ways. First, they were performed on patients who were not at immediate risk of death. Many other surgeries were obviously necessary: removing a mangled limb or repairing an open wound. Even the relatively common procedure of lithotomy (removal of bladder and kidney stones) mostly happened to patients who were in danger of dying, having lost their ability to urinate. By contrast, cancer patients chose present agony to avoid a drawn-out demise months or years down the line. Secondly, mastectomy was one of the most invasive and dangerous operations a surgeon could attempt, or a patient could undergo. To begin, the surgeons – commonly known as 'chirurgeons' – would cut into or around the breast with either a razor or a sharp thin wire and lift away a chunk of tissue. Some surgeons made a noose around the affected area so that it could be lifted away more easily (this was why Clark and Linch brought their 'waxed thread').[2] Then, the areas around where the tumour had been had to be excised either with a knife or with a red-hot iron which burned away the flesh. Surgeons knew that 'a Canker once cut often doth come again', and there was seldom a second chance at an operation.[3] Every part of the tumour had to be eradicated. The final task then was to stop the bleeding, since this was what could kill a patient before the operation was even finished. Small wounds could be stitched up, but in an operation like this the blood vessels were too many and too large to suture. Some physicians used boiling turpentine oil or hot irons to cauterise the wound, though they 'make the patient tremble'.[4] Others made do with styptic powders meant to slow blood loss, alongside liberal application of wads of cloth called 'pledgets'.

Drawing of a mastectomy. Attributed to a Dutch artist, seventeenth century. The Wellcome Collection.

Through all this, the patient had minimal, if any, pain relief. For milder hurts, medical practitioners gave medicine containing willow bark (this contains salicin, the active ingredient in aspirin) and other anti-inflammatory herbs such as plantain and cloves. Physician Alexander Read wrote that he also employed 'lily, the henbane, hemlock, the deadly night-shade, mandrake' and 'the apple of Peru' (a distant member of the potato family) in lotions to 'stupefy the part'.[5] All these substances are toxic in anything but small doses, and probably had a mild effect because they induced drowsiness. But the only substance with power to relieve intense pain in this period was opium, or its more commonly used tincture, laudanum. This wonder drug was the strongest in the physician's arsenal, often used in the palliative care of terminally ill patients. Surgeon John Woodall wrote that he used laudanum when patients were 'at death's door, or almost mad with the vehemency of the [pain], this precious medicine giveth ease presently, yea, and quiet sleep'.[6] Though it seemed like the answer to countless prayers, however, it came with such dangers that it was seldom used on anyone who was not already dying. Opiates were known to be addictive; Ward noted in his diary that 'I have heard of a woman that used constantly to take opium and could not sleep without it: she took to the quantity of 3 grains every night, and she had so used herself to it that she could take a great deal more'.[7] They were also potentially deadly, and in the quantities needed to anaesthetise a patient for surgery, could cause catastrophically low blood pressure, heart arrhythmia and coma. Writing about castration operations in the eighteenth century, French diplomat Charles Ancillon noted coolly that when continental doctors gelded young boys to create castrato singers, 'Sometimes they used to give a certain quantity of opium to the persons designed for castration, whom they cut while they were in their dead sleep…most of those that had been cut after this manner, died by this narcotic.'[8]

Though it seemed tantalisingly close, effective anaesthesia would not be invented until the nineteenth century, two hundred years after Mrs Townsend's mastectomy. In the meantime, the very thought of an operation struck fear into patients, and many surgeons were traumatised by the haunting screams of those under the knife. One instructional textbook advised young surgeons that women undergoing mastectomy might 'shriek and cry in a manner so terrible, as is sufficient to shock and confuse the most intrepid surgeon'. Regardless, the operator had to 'equip himself…as if he was deaf to the moving groans, and piercing shrieks, of the tortur'd patient'.[9] They needed to harden their hearts and hasten their hands, to practice as much as possible on senseless corpses and then try to think of living patients as pieces of meat. While the temptation was always to try and rush through the procedure, however, this had its own hazards. In 1684, the fashionable Swiss physician Théophile Bonet recorded, 'I saw a cancer so quickly cut from the breast that in the time of the operation I was scarcely able to speak three words: But when a cancer is come to a great bigness, then this speedy cutting is not proper…a great hole is made…and thereby the patient is more weakened.'[10]

It is hard now to say how many patients survived the horror of their surgeries, because surgeons had an infuriating habit of completely neglecting to mention the outcomes of the operations they depicted. Nobody wants to count their failures, after all. Nonetheless, the clues we have paint a grim picture. In eighteenth-century surgeon Daniel Turner's book about tumour operations, he provided details which showed around thirty percent of patients died under his care.[11] Turner was a skilled operator, and his notes included many simpler lumpectomies, as well as more serious cases. In 1639, John Woodall trumpeted it as a great credit to his techniques that during his long career as a surgeon in St Bartholomew's Hospital, amputating 'many more than one hundred of legs and arms, besides very many hands, and fingers', 'not above four of each twenty' of his patients had died while in his care, while John Ward recollected with admiration that 'Mr. Day hath amputated 5 arms 3 legs and somewhat else… and but 2 of all those died'.[12] Based on these snippets, we can guess that serious operations like mastectomies and limb amputations had a mortality rate of somewhere between twenty-five and fifty percent. The odds were appalling, but patients still pressed their surgeons to operate, for as Pierre Dionis bluntly told one patient, there was 'no other choice, but either that operation or death'.[13] What was more, just enough women recovered from this ordeal to give hope to their peers. Among the short notices of deaths and illnesses that appeared in seventeenth-century newspapers were tucked away near-miraculous stories of women like 'the lady of Sir Challenor Ogle', 'the wife of Mr Lee', and 'Lady Betty Hastings', who survived their operations and were, in the laconic terms of the newssheet, 'in a fair way of doing well'.[14] Surgeons and their patients were brave, desperate, sometimes reckless, but above all, hopeful.

***

One hundred and fifty years before Mrs Townsend's operation, while shrewd old Thomas Linacre was working on King Henry VIII to establish his College of Physicians, humbler men were watching carefully. These were the barber-surgeons, the donkeys turning the mill-wheel of the medical world; at least, that was how physicians perceived them. They did the grunt work other practitioners were too rarefied or afraid to touch: lancing boils, redressing hernias and fistulas, letting blood, setting fractures. Unlike physicians, their training was practical, not scholarly, and they learned on the job as apprentices. They saw how the physicians' new status elevated them in the eyes of polite society, and they wanted a piece of the pie.

They were about to strike it lucky. Sometime in the 1520s, Henry VIII found himself away from home and in need of a surgeon. He was making a royal progress – essentially a road trip with extra pomp and circumstance – through Kent, probably on his way to Canterbury, when he was seized by pain from a leg ulcer.* Such an ulcer could have been caused by any number of things, from old unhealed injuries to diabetes to varicose veins to syphilis, but it was debilitating for Henry, who, still in his thirties, prided himself on his riding, dancing and sporting prowess. Henry's attendants decided that as this was an external condition, it was best treated by a barber-surgeon rather than one of the king's physicians, and they summoned a local man of renown: Thomas Vicary.

A youthful but nonetheless experienced surgeon, Vicary would have treated plenty of ulcers before. They were a frequent complaint in this period, made common by poor treatment of wounds, rampant infectious diseases and long hours on horseback in heavy, uncomfortable clothes. He might still have quailed at the unpleasant treatment he was about to dish out to the most powerful man in the country. Peter Lowe, who would later be Vicary's colleague, described ulcers as 'horrible to look on, pale coloured, evil savoured…most sordid and stinking'.[15] They were painful, and they had to be treated in ways that were even more agonising, cutting through the dead flesh and into the sensitive living edges to encourage healthy tissue to grow. Sometimes ulcers ate deep into the body, and surgeons had to use probes to extract lumps of rotten flesh and bone. More fortunate patients might only need their sores dressed with ointments made with honey or egg whites. Whatever Vicary did to Henry's leg, Henry

---

* Henry was plagued by problems with his leg for the rest of his life, which was a major factor in his becoming increasingly obese and ill-tempered in his later years.

was grateful for it, for it was later said that Vicary 'was at first but a mean practiser in Maidstone...that had gained his knowledge by experience, until the King advanced him for curing his sore leg'.[16] By 1528, Vicary was employed as one of the king's surgeons, earning £20 per year, and by 1530, he was 'Sergeant-Surgeon', the highest-ranked surgeon in the royal household.[17]

Vicary had been catapulted to a position he could barely have dreamed of. Yet his pay was still modest compared to that of physicians, and he was still sneered at by some as a second-class medical practitioner who had 'gained his knowledge by experience' rather than by scholarship. Newlywed and more ambitious than ever, he saw how the College of Physicians had helped Linacre and his colleagues to advance. A similar college for surgeons was impossible, since they could claim no academic credentials, but Vicary and his colleagues persuaded Henry to grant the livery for a Company of Barber-Surgeons, which would unite the two professions under one banner while also establishing the boundaries of each. In centuries past, barbers had assisted monks in administering medical care. The alliance made sense; monks were not supposed to shed blood and barbers were handy with a razor (Vicary himself was a high-up member of the Company of Barbers). Now, though, that medical care was more secular and the great monasteries dissolved, Vicary thought it was time that the law caught up. Under the rules of the company, no barber was now to perform surgery, and no surgeon could shave or cut hair, though both groups could still pull teeth. In their union, surgeons would take precedence, and Vicary, the 'mean practiser' from Maidstone, would sit comfortably at the top of the tree.

The great event was commemorated in a picture by the most fashionable artist of the age, Hans Holbein the Younger, completed shortly before his death in 1543. As always, King Henry dominates the scene. Decked in ermine fur and silks, sporting his usual impressive codpiece, he wields the sceptre of monarchical power in one hand and in the other holds out the barber-surgeons' charter. Gazing up at him, like a dour choir of angels beholding their God, are surgeons, barbers, physicians and an apothecary, all wearing the sombre black robes and caps of their profession. On one side of Henry is Thomas Vicary, then the king's other surgeons, Sir John Ayliffe, James Monforde and Richard Ferris, and his barbers, Nicholas Simpson, Edmund Harman and John Penn, plus other men who would go on to positions of power in the company. On the other, looking less than thrilled about the new arrangement, are the king's physicians Dr John Chambers and Sir William Butts, plus his apothecary Thomas Alsop.

Notably absent from this painting – or any contemporary picture of surgery – were women. As surgeons sought a more professional status, they feared that 'tattling old wives' would contaminate the high-minded new company with 'the puddle of gross ignorance'.[18] But the fact that in 1630, the company's assistants had to officially warn their anatomy lecturers not to 'gather any money of any woman practiser', shows that women, who had long been practising all kinds of medicine without feeling the need for official say-so, felt differently.[19] As well as learned women like Grace Mildmay and Margaret Hoby, who would turn their hand to surgery when it was needed, there were many more wives, widows and daughters of male surgeons who soaked up medical knowledge, helped out with operations and even did some business of their own. Disputes between surgeons' wives and surgical apprentices show that the former were considered well able to train the latter, especially when the master of the

Henry VIII handing over a charter to Thomas Vicary, commemorating the joining of the Barbers and Surgeons Guilds, 1541. Oil painting by Hans Holbein the Younger. The Worshipful Company of Barbers/Bridgeman Images.

house was away. Sometimes they also assisted in unexpected ways. Wilhelm Fabry remembered how during a thigh amputation, his patient had begun to 'roar' so ferociously that his helpers ran away 'except only my eldest son, who was then but little, and to whom I had committed the holding of his thigh, for form only'. Fabry and the patient 'had been in extreme danger' were it not for his heavily pregnant wife, who came running from the next room and held down the struggling man.[20] The company was evidently reluctant to get in the way of such formidable characters, for though women were seldom licensed as surgeons, some widows were allowed to carry on their late husbands' trade. They kept a low profile, but women like 'Mrs Withers', 'Widow Moore' and 'Anne Hubbard' were no less important to their communities than their male counterparts.[21]

The establishment of the Company of Barber-Surgeons was a personal victory for Vicary, but it was just one link in the chain of events that would transform English surgery to rank among the best in the world. The company's new status meant that its members were for the first time allowed to put on human dissections for the education of their members, and were granted four bodies each year fresh from the 'Tyburn Tree', the infamous gallows near to what is now Marble Arch.[22] Between the physicians, the surgeons, city officials keen to display the corpses of wrongdoers and families wanting to bury their relatives, hanged bodies were in hot demand, but it was sensibly decreed that the surgeons' need was greatest. After all, much of their work involved structures which were delicate and had to be discerned by touch. Among the most common major surgeries was extraction of a bladder or kidney stone in a procedure known as lithotomy. This was a 'great and dangerous operation' in which 'life and death do so wrestle together, that no man can tell which of them will have the victory,' but it was also a rapid one.[23] Surgeons aimed to finish the procedure in no more than a few minutes, cutting through the perineum and bladder. If they did not have a good grasp of anatomy, both literally and figuratively, then they could leave a patient doubly incontinent. Samuel Pepys was so glad to emerge unscathed from his lithotomy in 1658 that he celebrated the day every year by having a slap-up dinner to which he invited the woman in whose house he'd had the operation. When he became rich, he commissioned an ornate box to house the stone that had caused so much trouble.*

* * *

* Pepys claimed that his stone had been the size of a tennis ball. Tennis balls were smaller then than now, but still measured about 2.5 inches across.

Cutting up the dead was one thing, but as one surgeon reminded his students, skill 'is not gotten with sitting on a cushion at home'.[24] To truly hone their craft, surgeons needed to practise on the living. The Renaissance supplied that opportunity on an unprecedented scale. During the sixteenth and seventeenth centuries, England was almost constantly at war, often entangled in several conflicts at once. Henry VIII's break with the Catholic Church made it open season for Catholic powers to attack England, and for their part, the Tudor dynasty attended every fight, anxious to secure their precarious political position. For eighteen years, from 1585 to her death in 1603, Elizabeth I waged war with Spain, while at the same time, more than twenty thousand English troops were busy quashing the resistance of Irish nobility to her plans to impose Protestant rule over their island. Through the early part of the seventeenth century, England supported what is now Belgium, Luxembourg and the Netherlands against Spanish rule, only to become later embroiled in a series of bloody sea battles with the new Dutch Republic they had helped to create. As if fighting with most of Europe was not enough, the English turned on each other in a vicious civil war: Roundheads against Cavaliers, village against village, even families torn asunder by competing political ideologies.

This war machine was fuelled by human bodies. During Elizabeth's reign, around fifty-five percent of all men aged between eighteen and thirty-nine would have spent time in the armed forces in one form or another.[25] To supply the voracious appetite of the navy for new sailors, armed 'press gangs' could board merchant vessels and snatch crew members, while the 1597 Act of Vagrancy allowed for young men of 'disrepute' – basically, anyone found looking idle – to be forced into service. The numbers of men 'pressed' climbed from 4,835 in 1597 to 7,300 in only the first half of 1599.[26] More commonly, though, would-be recruits were simply lured in by the promise of travel, excitement and free beer. Parties consisting of several high-ranking military personnel and a couple of privates would tour market towns waving a flag and beating a drum, offering people who might never have left their hometown before a life of adventure and a signing bonus. And as the century wore on and printing became ever cheaper, low-cost song sheets provided lyrics for ballads – story-telling songs set to popular tunes – that glorified military life. The 1673 *News from the Camp*, for instance, promised through the time-honoured medium of terrible song lyrics that

> Where e're our Prince pleases our arms to employ,
> We'll follow our leaders with courage and joy,
> By land, or by sea, whether battel or siege,
> We'll accomplish all things that our duties obleige.[27]

What propaganda like this failed to mention was the terrible cost of military service. War had always been a nasty business. As much as actual combat, soldiers suffered from filthy living conditions, bad weather and starvation. Half the men Elizabeth sent to fight in Ireland died there, many far from the battlefield in cold and unsanitary encampments where hypothermia and dysentery ran riot. Later, in the sixteenth century, there were new factors at play as armies capitalised on the deadly potential of firearms. In a book about 'weapons of fire' written in 1592, ex-mercenary Humfrey Barwick described how guns had changed warfare forever. 'I never saw any slain outright with an arrow', he recalled, 'but with harquebuze [arquebus, a large handgun] and pistol shot, I have been at [battles] several times, where 20,000 hath been slain outright, besides many wounded and maimed.'[28] Most prized among the new weapons were muskets, which Barwick assured readers could kill the best-armoured men at two hundred yards and the unarmoured at six hundred yards.[29] They would only become more deadly in the seventeenth century, when the development of the flintlock did away with the need to light one's gunpowder with a match, and so allowed guns to be fired twice as rapidly. At sea, the main danger was from cannon shot, which sent deadly high-velocity splinters flying through crowded, gunpowder-packed ships.

In 1617, an adventure story was published that showed just how dangerous the seas could be, and how important a role ships' surgeons played. *The Dolphin's Danger and Deliverance* related the true (or true-ish) account of the *Dolphin*, a small English trading vessel caught up in the ongoing war for dominance over valuable trade routes. The captain, Edward Nichols, told how they'd been on their way back from a trip to Zante and sailing close to Sardinia when they spied six Turkish men-of-war, huge ships with three masts, two hundred crew apiece and bristling with artillery. There was no escape, and as the ships closed in, Nichols gathered his crew to muster their spirits. 'We must be men or slaves', he urged. 'Die with me, or if you will not, I by God's grace will die with you.'

Nichols's prediction came true for many of his men, as seven of the thirty-six-strong crew perished during the skirmish. They included William Sweete, a trumpeter who continued sounding his instrument after having one arm blasted off, only stopping when another shot removed his remaining arm 'trumpet and all'. Walter

Penrose, William Russell, John Sands, David Fause, Thomas Shepheard and 'a boy', Benjamin Cornelius, were likewise shot dead. Within a few days, four more had succumbed to wounds including gunpowder burns, having a shoulder blade shot off, and being blinded. Of the remainder, five were 'maimed' but alive, including men who had lost an eye or been shot in the thigh, and one sturdy individual who had twelve separate shot and splinter wounds.[30] As well as listing the dead, Nichols gave a special mention to the ship's surgeon, Robert Grove, who during the battle with the Turks found that 'a ball of wildfire' fell into the basin he was using to rinse off his surgical instruments. When he tried to throw the incendiary device back, it hit the deck, and as Nichols recorded, would have set fire to the ship if Grove had not immediately fallen upon it and smothered the flames with his own body.

Whether at sea or on land, the field of battle was the crucible in which great surgeons were forged, and new surgical techniques were put to the test. Even relatively modest ships like the *Dolphin* had a surgeon on board, and back in London, men like the apothecary John Houghton acted as recruitment agents, promising that 'if any masters of ships want chyrurgeons or chyrurgeons [want] voyages, I'll strive to help them'.[31] These environments were physically and mentally tough for young medics, but also unique in the sheer numbers of bodies to practise on. On long campaigns, soldiers and sailors needed the surgeon's help to ease the effects of scurvy, venereal diseases, malnutrition and frostbite. When the enemy was finally encountered, the surgeon would see complex shrapnel injuries, open wounds of the abdomen, punctured intestines or lungs, brain injuries and open fractures. Whereas a surgeon might do a handful of amputations in a year in civilian practice, in combat they came thick and fast.

In 1686, John Moyle wrote a book aimed specifically at young surgeons preparing to go to sea. He counselled his readers that before an encounter with enemy ships, they should prepare two chests full of water. One chest was to dip the operating instruments into between patients, as close as early modern surgeons came to disinfecting their tools. The other was 'to throw amputated limbs into until there is conveniency to heave them over-board'.[32] Every time Moyle's ship entered battle, he expected to be sawing off so many arms and legs that he would need a designated place to stow them.

Military life was full of hazard and hardship, and military medicine was little better. Yet this turbulent era was also a catalyst. When resources were scarce and time was short, the demands of battlefield surgery forced medics to innovate, and new techniques found their way into civilian practice. In fact, the foundations of this surgical revolution had been laid a century earlier, by a military medic who made his name in the battlefields of France.

The trailblazer: Ambroise Paré

Ambroise Paré was the man whose work had by far the greatest influence on English surgery in the seventeenth century. Yet he was neither English, nor lived to see the years in which his ideas would have most impact. Born in 1510, he was the son of a cabinet-maker from Bourg-Hersent, a quiet village nearly two hundred miles west of Paris. Paré freely admitted that he never learned classical languages like most well-to-do boys of his time, and later in life he would have to fight with medical authorities and gain the backing of the king to be allowed to publish his works in French, not Latin. He did, however, catch an interest in medicine from watching his older brother Jehan, a surgeon in the town of Vitré to the west. By 1532, he was apprenticed to a barber-surgeon in Paris and soon was seeing poor patients in the dingy surroundings of the Hôtel-Dieu, the city's oldest hospital.

Training here was a mixed blessing. On one hand, the Hôtel-Dieu was an unhygienic, depressing place, dank with foul air from the nearby Seine. Even two hundred years later, one of its surgeons would attest that it was 'the most unhealthy and uncomfortable of all hospitals', with a quarter of patients dying in its dubious care.[33] On the other, it was a good place for a young medic to cut his teeth, with lots of cases and minimal supervision. For Paré, it beat the old-fashioned education and expensive exams offered by the nearby University of Paris. His course as a pioneering surgeon was set, and in 1536, when war broke out between the French King François I and the Holy Roman Emperor and King of Spain Charles V, he saw his chance.

The realities of life as a novice military surgeon were a shock to young Paré. In fact, he later admitted that he had been way out of his depth: 'I will tell the truth, I was not very expert at that time in matters of chirurgery; neither was I used to dress[ing] wounds made by gunshot.'[34] He had to catch up fast; as scores of wounded flooded into his makeshift operating room, Paré was equally revolted and rapt to discover how musket shot obliterated limbs beyond repair as 'the bullet by its great force crushes and breaks not only the bones that it touches but also those that are far from them'.[35] What he was observing was the effect of the lead casting process, which created a large (two-centimetre) projectile with an uneven surface.* When they met human flesh, they deformed and refracted, winding a drunken, reeling path through the body. To make matters worse, many physicians at the time believed that gunshot was inherently poisonous, which was why the wounds left black marks on the skin and so often became infected (later surgeons worked out that the problem was that the projectile drove fragments of cloth and dirt into the wound).

Belief in the toxicity of gunshot was such that Paré reported watching seasoned soldiers mix gunpowder with their wine to try and gain immunity from the poison.[36] The standard treatment was to cauterise gunshot wounds with boiling oil, a practice which had been popularised ('popular' being a *very* relative term here) by the Italian surgeon Giovanni da Vigo in the fifteenth century. On his first military job, however, the battle was so bloody that Paré ran out of oil. Ever resourceful, he dressed the wounds using a mixture of egg yolks, oil of roses and turpentine instead, but he feared the consequences. Later, he recalled, 'I could not sleep all that night…I feared that the next day I should find them dead, or at the point of death by the poison of the wound, whom I had not dressed with the scalding oil.' Yet when morning finally came, he rose early to a surprise:

> I visited my patients, and beyond expectation, I found such as I had dressed with a digestive only, free from vehemency of pain to have had good rest, and that

* Muskets were relatively new weapons, having developed in the early sixteenth century from the smaller 'arquebus' (though the terms were often used interchangeably).

> their wounds were not inflamed, nor tumified [swollen]; but on the contrary the others that were burnt with the scalding oil were feverish, tormented with much pain, and the parts about their wounds were swollen. When I had many times tried this in diverse others, I thought thus much, that neither I nor any other should ever cauterize any wounded with gun-shot.[37]

Through sheer luck, Paré had proved that the hot oil method was in fact *more* harmful than regular dressings, a truly momentous finding. Admittedly, this revelation was followed by some missteps. After the success of his makeshift ointment, the young operator set about trying other concoctions, and was particularly delighted when he persuaded a renowned surgeon to give him his top-secret recipe – a mixture which included two new-born puppies boiled in oil with a pound of earthworms, all mixed with turpentine.[38] Nevertheless, he had established the principle that boiling oil cautery was a poor way to treat dangerous blood loss, and he told anyone who would listen that he now favoured the use of ligatures on severed arteries, a difficult technique which was not widely used for amputations. This was to be a vital step in the surgical advancements of the next two hundred years, simply by keeping patients alive for long enough to endure bolder, more complex procedures. Though Paré's professional rivals sought to discredit his approach – fellow surgeon Etienne Gourmelen called it 'arrogant' and 'indiscreet' – his patients welcomed anything that lessened their suffering.[39] At the French siege of Danvilliers in 1552, Paré amputated a man's leg which had been shattered by gunshot, using ligatures and ointment instead of hot irons and oil. Though the man still had to endure his leg being sawed off without anaesthetic, Paré wrote that when he got back to Paris '[he] was content, saying that he escaped good cheap, not to have been miserably burnt'. For his part, Paré concluded, 'I dressed him, and God cured him.'[40]

Paré dipped in and out of military surgery throughout his career, also making advances in obstetrics and general medicine. Where the Parisian medical elite were hamstrung by their devotion to the ideas they had inherited from classical texts, he emphasised experimentation and practical experience. In the process, he met Andreas Vesalius, the man who had revolutionised the study of anatomy, on at least one and probably two occasions.

The first was in 1553, when Paré was acting as military medic for the French forces in the town of Hesdin. They were forced to surrender the town to the opposing side, that of the Holy Roman Empire, and Paré found himself in front of the emperor's chief surgeon – a man whose name is not given in surviving records, but was very likely Vesalius. The encounter was strange, and tense. With Paré was another member of the French forces who had been shot in the lung, and Vesalius asked Paré to explain the nature of the wound and his treatment for it, which he duly did. When the patient died, Vesalius, ever the anatomist, asked that Paré autopsy the body. Paré demurred, claiming that he was unworthy to operate in such company; Vesalius told him that if he refused, he might swiftly come to regret it. At the end of the encounter, Vesalius released his colleague-cum-hostage, but not before trying, unsuccessfully, to persuade him to join the empire.

The second encounter happened in 1559, when a fragile peace between France and Spain had finally been brokered. One condition of the deal was that King Philip II of Spain would marry Elizabeth, the thirteen-year-old daughter of the French King

Henry II, and the Duke of Savoy Emmanuel Philibert would marry Henry's sister Marguerite. To celebrate the occasion, a lavish three-day tournament was held, but during a joust on the final day, King Henry was struck in the face by the lance of the Count of Montgomery, a captain of the Scottish Guard. Pieces of the lance shattered at the weak spot just underneath the king's visor, penetrating his right eye and entering his brain. Paré – by then a fixture of the royal retinue – instantly began treating the stricken king, but keen to appear helpful, the Duke of Savoy also sent for the great Vesalius, then working in Belgium. Once again, the two giants of their profession met under fraught conditions, this time knowing that their success or failure might reverberate through European politics. Their combined expertise could not save the king. Henry died a few weeks later, aged just forty, and Vesalius wisely beat a hasty retreat to Brussels before anybody could make him scapegoat.[41]

From the field of battle to the royal court, Paré became known for his calm, unassuming manner and plain speaking. The combination of his character and expertise brought him into ever more rarefied circles; he served as master-surgeon to four French kings and the Queen Consort of France, the formidable matriarch Catherine de' Medici. But it was his early discoveries as a military surgeon that one day came to benefit him directly.

In May 1561, aged fifty-one, Paré was travelling with three fellow medics: Monsieur Nestor, a noted physician, and the surgeons Richard Hubert and Antoine Portail. They planned to visit a group of Minim monks known as 'bonshommes' (good men), in north-east France, but to do so they had to persuade their horses to get into a boat to cross a river. Impatient to get going, Paré hit his horse on the backside with a riding crop, at which the disgruntled animal kicked him hard enough to shatter both bones of the lower left leg. He recalled how as he reeled back from the kick, the jagged bones broke through his flesh, clothes and boot, causing 'such pain that it is not possible for a man…to endure greater without death'.[42] At one point, such a wound would have been treated with boiling oil, but now his companions rushed to the nearest house to make an ointment out of eggs, flour and soot from the fire. Though Paré's recovery was long and painful, the surgeons were able to extract the shards of bone and reduce the fracture, and he did not lose the leg as he had feared. At the end of his treatise on the subject, he prayed to God 'to send me death rather than to fall into [such an accident] again'.[43]

Though it was an ordeal he never wished to repeat, Paré's brush with death had unexpectedly positive consequences. Paré had always been interested in limb injuries, but now he was even more preoccupied with gangrene, amputation and disability. After all, what if *he* had been one of those 'poor cripples' to lose a limb, never to have walked or worked again? Increasingly, his thoughts turned to the innumerable men he had seen maimed on the battlefield, who returned home both physically broken and socially ostracised. Men who'd suffered injuries to the spine were permanently crook-backed, while those with damage to the penis might be left incontinent. The palate injuries that often resulted from shrapnel wounds meant a life without the power of speech, while the loss of an arm, leg, eye, ear or nose not only affected a person's ability to work, but their chances of simply appearing in public without being ridiculed. This was not a good time to be different.

Paré resolved that such people 'must be so trimmed and ordered, that they may come in a seemly manner into the company of others'.[44] As well as surgical procedures, he began to stuff his books with what we would now recognise as prostheses. Some were simple but effective. For a man who'd lost his lip in a sword fight, he recommended 'there be a lip of gold made for it, so shadowed and counterfeited, that it may not be much unlike in colour to the natural lip, and it must be fastened and tied to the hat or cap that the patient weareth on his head, that so it may remain stable and firm'.[45] When the eye was lost, the patient could wear a glass eye, or if the whole socket was destroyed, they might use a leather patch with an eye painted on, attached around the ear with a wire. New noses could be crafted 'of gold, silver, paper or linen cloths glued together', and 'so coloured, counterfeited and made both of fashion, figure and bigness, that it may as aptly as possible, resemble the natural nose: it must be bound or stayed with little threads...unto the hinder part of the head or the hat'.[46] For some, like the Danish astronomer Tycho Brahe, it became a signature look.

But while painted patches and gold noses were all very well for the face, the body required different kinds of help. Paré drew blueprints for basic trusses 'to amend the crookedness of the body' and small funnels to enable urination for 'those that have their yard [penis] cut off close to their bellies'.[47] To design his most sophisticated prostheses, however, he needed the help of a craftsman, a Parisian mechanic he nicknamed 'le petit Lorrain'. Inspired by the handiwork of this artisan, Paré produced detailed drawings of prosthetic arms, legs and hands, designed 'not only for the action of the amputated parts, but also for the[ir] beauty and ornament'.[48] For instance, the artificial hand he proposed contained small cogs and pulleys which allowed each finger to bend independently, and was plated with metal which could be

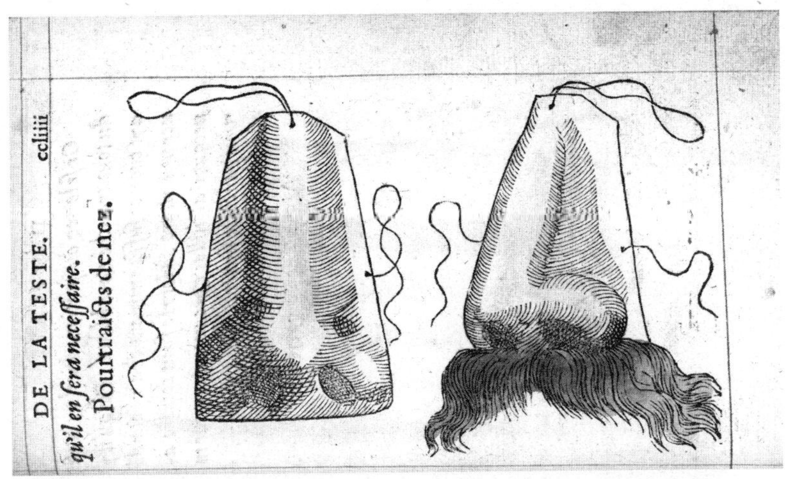

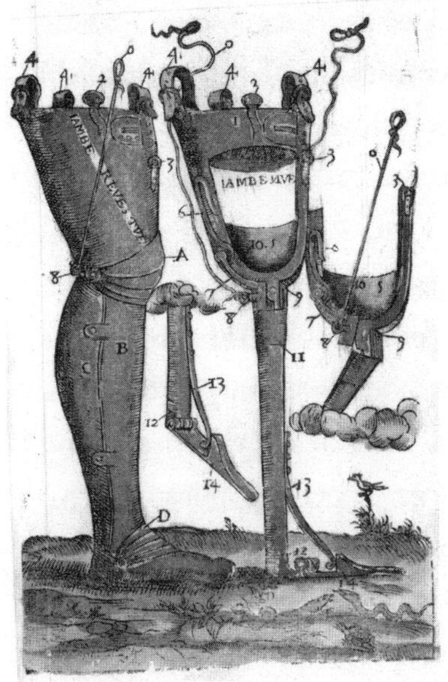

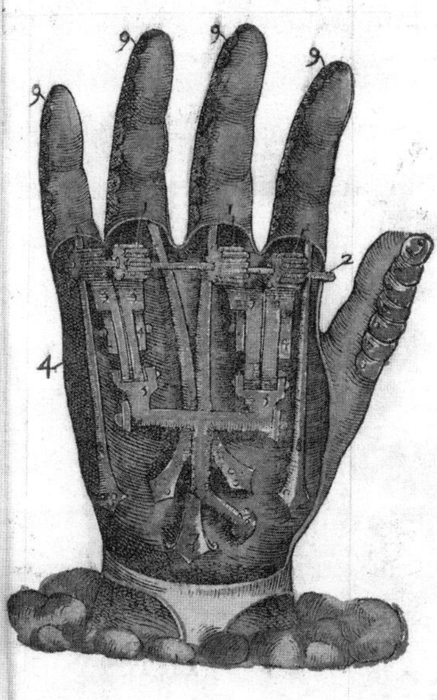

Blueprints for: 'a prosthetic nose (with or without moustache)'; 'a mechanised prosthetic leg'; 'a mechanised prosthetic hand'. From *Instrumenta chyrurgiae et icones anathomicae* by Ambroise Paré. The Wellcome Collection.

enamelled to match the patient's skin tone. A false leg used thin cords to allow the toe to drop and pick up with each step. Overall, the aim was to imagine prostheses with which amputees could 'perform the functions of going, standing and handling', and which were 'not only profitable for the necessity of the body, but also for the decency and comeliness thereof'.[49]

Most of Paré's prosthesis designs did not see the light of day, at least not in his own lifetime. Nonetheless, they would prove influential. Paré had always said that he was not interested in actually *making* false limbs, but rather, 'I have caused them to be portrayed or set down, that those that stand in need of such things, after the example of them, may cause some [black]smith, or such like workman to serve them in the like case.'[50] He would not live to see this, but in the decades following his death, a new character entered the medical stage: the prosthesis maker.

Perhaps more importantly, though, Paré's work prompted surgeons to reconsider what they could achieve in trauma surgery, and what kind of existence was possible for people after life-changing injury. In 1634, forty-four years after his death, *The Workes of that Famous Chirurgeon Ambrose Parey* were translated into English and printed as a large, expensive book. Twelve hundred pages with a fine frontispiece and many illustrations, it was commissioned by Thomas Cotes, a man with an impeccable pedigree in the publishing world. As a lad, he'd been apprenticed to William Jaggard, who printed Shakespeare's First Folio, and only two years earlier he had produced the Second Folio, having inherited Jaggard's catalogue and business.

Sure enough, Cotes's instincts proved correct. *The Workes* swiftly became required reading for any self-respecting surgeon, and prompted many to reconsider the life awaiting patients after they left their care. Contemporary surgeons could not agree on the best place to sever a damaged limb, whether to preserve as much as possible or cut it off near the knee or elbow to better accommodate a prosthesis. John Woodall had long counselled that parts of a limb which were 'guiltless' – that is, not gangrenous – should be preserved at all costs, 'for the nobleness of each member of man's body, and namely of the leg, is highly even in humanity to be tendered and regarded'.[51] His opinion was based equally on religious faith and bitter experience. He urged surgeons to impress upon their patients 'the eminent danger of death by the use thereof [of amputation]', adding that for the operator 'it is no small presumption to dismember the image of God'.[52] Though Paré likewise knew the perilousness of amputation, he prioritised function over preservation. He firmly believed that even if most of the patient's leg was sound, it was nonetheless best to cut it off just a few inches below the knee, so that 'he may the better go on a wooden leg'. He gave as proof the example of Captain Francis Clarke, who found that after the amputation of his foot at the ankle 'he was much wearied with the heavy and unprofitable burden of the rest of his leg'. Extraordinarily, Clarke requested a further amputation to remove the rest of the leg up to the knee, after which he used it 'with much more ease and facility than before'.[53]

Both surgeons and patients were soon convinced by Paré's argument, and cutting off limbs (especially legs) at the joint became the new orthodoxy. For example, the 1699 *Compleat Body of Chirurgical Operations* suggested that 'The leg must be cut off as near the knee as possible…for the more commodious carrying a wooden leg…On the contrary cut as little as may be off the arm, because it serves as an ornament and counterpoise to the body, and an artificial hand may be made to be useful in some cases.'[54] This simple change shows how each piece of the medical marketplace was connected. While prostheses were heavy and useless, there was not much point in cutting off limbs in a way that facilitated prosthesis use. Conversely, there was no incentive for people to develop better false limbs if amputees were not able to wear them. It took an exceptionally forward-thinking surgeon to break this impasse – one whose interests ranged far and wide, who had personally confronted the kind of life-changing injury experienced by many of his patients, and who was unafraid to break with tradition.

***

With the ambition they had learned from Paré, the credibility lent by company membership and the bitter experience gained from years of war, English surgeons transformed from the poor relations of learned physicians to artists of practical medicine. The lecture notes of one diligent seventeenth-century student tell us of the incredible range of operations he learned: trepanning, amputations, draining fluid from the abdomen, removing tumours, lithotomy, fixing cleft lips, tracheotomy, 'stitching the tendons', redressing hernias, an operation to fix phimosis (inability to retract the foreskin of the penis) and two ways to address fistulas.[55]

Nevertheless, we should not forget that to excel among Renaissance surgeons was still to be the best in a horrifying, bloody trade. The surgeon – his robes black to hide the gore that accompanied his work – still inspired fear wherever he appeared, and many ordinary people still mistrusted 'sawbones' just as they did physicians. The personal paperwork of seventeenth- and eighteenth-century England reveals the stories surgeons left out of their textbooks. They show good and bad, a marketplace that was expanding rapidly and a populace who demanded more from their medical practitioners. But in the newspapers of the time, surgeons like William Read, 'Her Majesty's Oculist', commissioned advertisements boasting about their success in turning children with cleft lips from 'monstrous spectacle' to 'not now in the least deformed', as well as 'couching' cataracts (dislodging the cloudy lens), curing cancers and removing 'wens' (growths).[56] In the north of England, a man who'd had his leg amputated had it buried in the local churchyard with a stone bearing the inscription:

> Here lies the Leg of Master Conder:
> But he's alive, and that's a Wonder.
> It was cut off by Dr. Johnson,
> The famousest Surgeon of the Nation.[57]

These were tales of noble-minded medics swooping in to save the day, laying hands on the sick and changing their lives. They helped inaugurate a narrative of surgeons as lone heroes which remains alive and well in some corners of the medical profession today. To see the other side of the story, we must turn to different sources. In the Middlesex assizes courtroom, the impoverished William Leeson begged the justices to grant him £3 a year in light of the fact that 'when your petitioner's leg was shot off, the surgeon cut out the hamstrings of his stump, which renders him incapable of walking the fourth part so well as any other man which hath had a *good* amputation'.[58] In a dispute between surgeon John Thorpe and his assistant Bartholomew Penny, Penny countered Thorpe's accusations of misconduct by alleging that the surgeon was not only incompetent, but half blind.[59] Even surgeons complained about other operators who they said flouted regulations and worked without proper training. William Fabry warned against the 'strowling mountebanks and imposters, [who] unadvisedly apply the knife to the body of man, like butchers to the brute beasts' and 'care not if they buy their skill with the deaths of a hundred common people, as I have heard it from their own mouths'.[60] And as always, the rivalry between physicians and surgeons raged, not helped by the latter's repeated attempts to tread on the physicians' toes by asking to be allowed to prescribe medicines.

These tales of misconduct, slander and sometimes outright villainy coloured public opinion at the time, and our modern ideas about Renaissance surgery still tend to dwell on its undeniable horrors. There was no humane way of operating on people before anaesthesia. Nonetheless, the reason that so many stories survive, both good and bad, is that this area of medical practice was exploding into life. Surgeons and other medical practitioners griped and sniped and took pot-shots at each other because they were jostling for position, competing for prestige, recognition and ultimately customers. The medical marketplace was now bulging at the seams – and there was much more still to come.

As for Mrs Townsend, the desperate chance she took in opting for a mastectomy did not ultimately pay off. The next time John Ward saw her was at her post-mortem, when with a 'Mr. Eedes', he opened her breast and found 'two porringers [porridge bowls] full' of a yellow substance, plus gristly scar tissue. Though Ward's diaries are vague on dates, this was only weeks or months after the operation which they'd hoped would save her life. Was the cancer already at work deeper in the body, or was the yellow substance pus from a deadly infection? Ward could not tell, for he did not have the equipment needed to delve further into the corpse.[61] However, his diary ensured that, unlike other patients, her fortitude and bravery were remembered. Whatever ultimately killed her, Mrs Townsend was prepared to risk everything for a chance at life.

A man composed of pharmaceutical equipment, surrounded by medicinal plants. Engraving by N. de Larmessin, 1695. The Wellcome Collection.

CHAPTER FOUR

Medicine for the Masses: Apothecaries

In the pit that faces the Globe Theatre's raised stage, playgoers shift from foot to foot. Lovers hold hands, women wring handkerchiefs and grown men hold their breath. Rapt by the action unfolding before them, an unlucky few have their purses cut. On stage, the lovelorn Romeo is learning of the apparent death of his Juliet. Exiled in Mantua, his thoughts turn immediately to his own death. Poison is to be the means, and he knows where to get it.

> I do remember an apothecary
> (And hereabouts he dwells) which late I noted
> In tattered weeds, with overwhelming brows
> Culling of simples. Meagre were his looks.
> Sharp misery had worn him to the bones.
> And in his needy shop a tortoise hung
> An alligator stuffed, and other skins
> Of ill-shaped fishes; and about his shelves,
> A beggarly account of empty boxes,
> Green earthen pots, bladders, and musty seeds,
> Remnants of packthread, and old cakes of roses
> Were thinly scattered to make up a show.
> Noting this penury, to myself I said
> 'An if a man did need a poison now,
> Whose sale is present death in Mantua,
> Here lives a caitiff wretch would sell it him.'

When Shakespeare wrote *Romeo and Juliet* in the early 1590s, he had never – as far as anybody knows – been to Mantua, or Verona, where the play's dramatic finale unfolds. Rome, Venice, Navarre and Ephesus were exotic names, far removed from the everyday bustle and dirt and seemingly endless rain of England. But while the bard might never have set foot on Italian soil, he would have browsed plenty of apothecaries' shops. Pungent with the mingled scents of hundreds of 'simples', or ingredients, they were a feast for the senses. Exquisitely decorated pots adorned the shelves, while perfumed steam and the harsh sound of pestle on mortar issued from the back room where medicines were concocted. Tobacco, newly arrived from the Americas, sat in jars behind the counter, facing the desiccated remains of long-dead creatures from distant climes. Most people in this period had never even seen pictures of a tortoise, or shark, or any of the other exotic specimens which apothecaries hung in their dimly lit shops to entertain curious customers as they waited. To enter an apothecary's shop was to step into another world.

With its everyday marvels, the apothecary's shop pulled off the magic trick of being both extraordinary and very commonplace. By the time Shakespeare put pen to paper, there was roughly one licensed apothecary for every two thousand people in London, and many more unlicensed ones. That made them ten-a-penny compared to physicians or surgeons, though still less numerous than the many domestic healers, nurses and itinerant medicine-sellers. Their shops flourished on the roads that led towards wealthy parts of London such as Westminster and St Paul's; in the parishes of St Benet and St Stephen Walbrook, close to the mighty cathedral, one in five householders were druggists or apothecaries. It was a good, steady way to make a living, since there would always be sick people. Unlike a physician's pricey services, the apothecary's wares were accessible to all but the very poorest; theriac, which was used for everything from malaria to indigestion, retailed from just one penny. Nonetheless, the sheer number of medicines they supplied allowed them to turn a decent profit. Between 28 August and 30 October 1633, for example, an unnamed apothecary supplied 'Mr Wimkells' with thirty-one items, ranging in price from 8d [pence] for a 'fume for the head' (probably a mixture to be inhaled to ease headaches) to 2s 6d for 'a bottle of diet' (a drink to help indigestion).[1]

The recipe for making an apothecary was tried and true. Take a callow youth of about age fourteen, round with puppy fat or lean and long-limbed. He should be quick-witted, able to read and write, preferably with a bit of Latin (though not essential). Subject him, over seven years, to gradually increasing heat. He needs a strong stomach, not only to prepare noxious medicines, but to let blood, assess sick patients and, occasionally, embalm corpses. If necessary, beat the youth vigorously. Apprentices are a rowdy bunch, prone to riot.* According to their temperament – and their treatment – they may become as beloved as a son or as loathed as a slave. Finally, a little sweetening goes a long way. Six times a year, anyone able to leave his shop for the day might bring along his apprentice and set off on a 'herbarizing', walking from St Paul's to the green spaces just outside the city. The purpose of these expeditions is allegedly to teach the next generation how to

* In 1595, the year in which *Romeo and Juliet* was first staged, over one thousand apprentices rioted in the streets around Tower Hill. They were protesting social inequality and in particular the price of food – the cost of flour nearly tripled between 1593 and 1597. In response, the city authorities imposed a curfew and temporarily closed the theatres.

recognise common herbs and flowers, but the main draw is the lunchtime beer and cakes.

On paper, the apothecary's role was a benign, even dull one, somewhere between a shopkeeper and a pharmacist. He doled out familiar potions and dutifully filled the prescriptions written by physicians according to the *Pharmacopoeia Londinensis*, a huge book compiled by the College of Physicians which told apothecaries what medicines they were allowed to make and exactly how to prepare them. The reality was much more colourful. Apothecaries evolved from the pepperers and spicers of medieval times, and when they began specialising in medicinal ingredients, they used their encyclopaedic knowledge to formulate their own medicines, as well as venturing into diagnosing and treating patients. Like the physicians and surgeons, apothecaries too gradually formalised their practice, campaigning to split from the Grocers' Company and be recognised in their own right. The struggle paid off in 1617, when the Society of Apothecaries was finally established. But the grocers were so furious about losing a quarter of their membership that James I had to step in to stop them unilaterally declaring a monopoly on everything from prunes to arsenic. Then there was the tedious wrestling for power between the physicians and the apothecaries. Physicians felt that they ought to have control over the drug trade; apothecaries, predictably, disagreed, leading to bitter disputes throughout the seventeenth century. But unlike other such disputes, this was more than just a war of words. The stakes of the debate had played out in spectacular fashion in the early 1600s, when an apothecary became embroiled in the most sensational criminal trial of the century, the notorious 'Overbury affair'.

* * *

The Overbury scandal started with friendship and romance and ended in murder. In the early years of the seventeenth century, Thomas Overbury, a well-educated lawyer, met Robert Carr, page to the Earl of Dunbar. The two men had much in common; both were in their early twenties, on the outskirts of courtly society and eager to rise above their small-time gentry origins. Overbury's prickly nature was mollified by Carr's easy joviality, while his shrewdness balanced Carr's naivety. They hit it off – Overbury became a friend, secretary and mentor to Carr, and together the two travelled to London to make their careers in the court of King James I. Carr had the advantage of being a good-looking young man at court with a king known to have a weakness for handsome courtiers. He swiftly became James's favourite companion, and he and Overbury began their ascent, Overbury being knighted and Carr becoming Viscount Rochester and a privy councillor to boot.

Everything was going to plan, until the moment Carr set his eyes, then heart (and possibly some other body parts), on Frances Howard. Frances was an auburn-haired, lily-skinned society beauty, forthright and intelligent, whose portrait showed her wearing a fashionably (yet still scandalously) low-cut gown. Unfortunately, she was also married to the Earl of Essex. Their union, contracted when both partners were still in their early teens, was deeply unhappy. Frances allegedly consulted the infamous astrologer-physician-alchemist-necromancer Simon Forman to obtain a potion that would extinguish her husband's already-flagging libido, while her husband regarded her with distrust and loathing. By contrast, Carr was infatuated, and the politically savvy Howards were keen to see their daughter leave the earl and marry this rising star.

There was just one problem. Overbury, who was known for being overbearing and arrogant, utterly disapproved of his friend's affair with such a 'filthy base woman' and wrote a long poem, circulated around the court, which very pointedly emphasised the sanctity of marriage and warned against lust.[2] Worse, he threatened to use his influence at court to prevent Frances getting the annulment she needed. He needed to be got out of the way, and the Howard family began to manoeuvre. Inch by inch, Carr, the king's darling, turned the monarch against Overbury, until the final straw came in 1612 when Overbury unwisely declined an offer (masterminded by the Howards) to become ambassador to Russia. James was furious and sent Overbury to the Tower to think about his priorities. Given that prisons at this time were a hotbed of disease, and the Tower little more than a fancy prison, nobody was too surprised when Overbury died after six months. His body, covered in 'diverse blains and blisters', was unceremoniously buried, and a rumour circulated that he had perished of venereal disease.[3]

With Overbury out of the way, Frances Howard was free to dissolve her marriage, resulting in a bitter trial where she claimed that her husband was impotent; he countered that he could feel 'motions and provocations of the flesh' perfectly well with any woman except her.[4] To help the cause, Frances submitted herself for medical examination to prove that she was still 'a virgin and uncorrupted', but since she appeared for the inspection in a thick veil, it was widely rumoured that she had paid another woman to take her place.[5] In any case, the annulment was granted, Carr and Frances married, and the whole affair seemed to be over. John Donne, the melancholic preacher-poet, even penned a lengthy poem in praise of the new union, an *epithalamion* (he had

The Portraitures of Robert, Earl of Somerset and Lady Frances Howard by Renold Elstrack, 1618. National Gallery of Art, Washington. D. C.

no love for the couple, but Carr was important to him, regularly doling out cash to Donne, albeit whilst failing to help him in his main aim of joining the priesthood). Yet the scandal would not die. An anonymous verse, quite different to Donne's, started to circulate:

> A page, a knight, a viscount and an Earl
> All four were wedded to one lustful girl
> A match well made, for she was likewise four
> A wife, a witch, a murderer, and a whore.[6]

The page, knight, viscount and earl of the poem was obviously Robert Carr, whose ascent through courtly ranks had made him powerful enemies. The less flattering quartet of titles was, of course, Frances. Public opinion had turned, and Carr, ousted from the king's affections by the dashing new courtier George Villiers, could no longer depend on royal protection. In 1615, trials began to establish the true cause of Overbury's death.

Enter the apothecary. The court heard that, in fact, Overbury's death had been anything but natural. His jailer, Richard Weston, was friends with a woman called Anne Turner, a widow who made ends meet by selling a coloured starch that stained lace ruffs and cuffs a fashionable yellow.* She in turn was close friends with Frances, and together they conspired to send Overbury numerous 'gifts' of tarts and jellies – each one, of course, containing a small

---

* It was not a coincidence that Weston was guarding Overbury. Frances's great-uncle, the Earl of Northampton, had already arranged that a friend of the family, Sir Gervase Helwys, should be appointed Lord Lieutenant of the Tower. Unlike his predecessor, Helwys was open to bribes, and took £2,000 to ensure that Weston was put in charge of Overbury.

amount of poison. The source of the poison? An apothecary named James Franklin, whose dubious credentials included the supposed ability to raise devils. Ominously, his own wife had also died under suspicious circumstances a few years earlier. According to the philosopher Francis Bacon, who wrote a pamphlet about the trials, Frances and Turner went to Franklin and asked him for a substance that would kill 'by little and little'.[7] 'Bewitched' by Frances, and swayed by a substantial bribe, Franklin obliged, supplying the pair with a liquid which, when administered to a cat, caused it to expire, 'crying pitifully for two days'.[8]

Overbury soon began to sicken. In addition to the mystery liquid, his meals were being seasoned with mercury and powdered arsenic slipped into his table salt.[9] In fact, Franklin later claimed that for a time virtually everything Overbury had eaten had been laced with some poison or other.[10] Yet he was, stubbornly, clinging to life. Frances and Turner returned to Franklin, and asked for an enema to finish the job, which was delivered to Overbury in the guise of medicine for his mysterious 'illness'. This time, it was effective, and the prisoner died in agony.

Under interrogation – perhaps torture – Franklin confessed to everything he had done and probably some things he had not. His chief crime was supplying poisons including mercury, arsenic, powder of diamonds (ground glass), cantharides (a type of poisonous beetle called 'Spanish fly') and 'great spiders', which were thought to be deadly if eaten. He protested that he had not known what Frances planned to do with the poisons once she had them, but it was no good. A jury took less than fifteen minutes to find him guilty, and he was sentenced to death along with Turner and Weston. Frances, however, applied her considerable wit and charm to the court, and instead of being hanged, she and her new husband were sentenced to reside indefinitely in one of the Tower's more comfortable celebrity suites, eventually receiving a pardon in 1622.[11]

The Overbury trial would shape the course of medicine in the seventeenth century. It was used by both physicians and apothecaries as evidence for their case. The apothecaries said that Franklin's actions showed the need for a proper organisation which could regulate and keep records of everyone selling medicines – that is, a self-governed Society of Apothecaries. The physicians said that it demonstrated that only they could be trusted with the important job of monitoring the drugs trade, and they should be given unfettered access to and control over apothecaries' shops. The flames of the dispute were enthusiastically fanned by popular ballads which claimed that Franklin was not just a common crook but a would-be sorcerer, 'rais'd by the black art'.[12] The link between chemistry, alchemy and sorcery was complex and contentious. In the trial of Frances and her co-conspirators, it was claimed that she had first been helped in her scheming by the notorious Simon Forman, and that his assistance had been of the supernatural, perhaps even demonic, kind. Forman's wife deposed that the pair spent hours 'locked up in his study together', and the court was shown what looked to be enchantments written on parchment.* One had what appeared to be a small piece of human skin fastened to it.[13] Even among less shady characters (and virtually anyone was less shady than Forman), herbalism

* Forman also apparently had a piece of paper detailing 'what Ladies loved what Lords in the court', but it was not allowed to be read out – perhaps because the names of some of the justices appeared on it (Bacon 51).

and astrology often went hand in hand. Nicholas Culpeper, for instance, was both the most famous apothecary of the seventeenth century and an astrologer who occasionally predicted the end of the world.

The trial also demonstrated how the drug trade itself was changing. Advances in chemistry, combined with the discovery of the New World and increased trade with the Far East, meant that medical practitioners had access to substances with a potency their forbears could only have dreamed of. Most of Franklin's poisons – barring the spiders and the 'powder of diamonds' – had legitimate medical uses. As we saw in the case of Henry More's niece, arsenic was used to eat away tumours. Cantharides, or Spanish fly, was made from ground-up green 'blister beetles', found 'beyond sea' (here meaning southern Europe and Asia). It was known to be highly caustic, and was sometimes used in remedies to reduce tumours or provoke urine – though it was unclear how the latter supposedly worked. In 1694, a London immigrant physician named Joannes Groenevelt was dragged in front of the College of Physicians and imprisoned for allegedly having permanently disabled a woman by giving her pills made of Spanish fly. They had been meant to cure his patient Suzanna Withall of an unnamed 'hurt got in labour', but instead gave her racking pains and rendered her unable to walk.[14]

At the Overbury trial, four justices heard that Franklin had 'killed, poisoned and murthered' his victim primarily using mercury.[15] Dissolved in aqua fortis (nitric acid), mercury formed white salts that could easily be added to pills and drinks, or mixed into food. It would have been a staple of the apothecary's cabinet, because as the surgeon John Woodall put it:

> The perfect cure proceeds from thee,
> for Pox, for Gout, for Leprosie,
> For scabs, for itch, of any sort,
> These cures with thee are but a sport.[16]

'Pox' was the contemporary name for what we would now call syphilis, which ran rampant among citizens of all European nations, each passing the naming buck onto the next (the Spanish pox, the Neapolitan pox, the French pox…and so on). It was spread principally through sex with infected partners, often sex workers – a dose of the pox was virtually standard issue for soldiers on campaign.* Coming in stages, it initially presented with mild symptoms such as blistering around the genitals, which went away of their own accord after a few weeks. Afterwards, a second wave brought a faint red rash, accompanied with headaches, fever, muscle pain, swollen glands and a sore throat, which again tended to disappear with or without treatment. Only a small number of infected patients went on to develop tertiary syphilis, which could bring about blindness, deafness and dementia. Most shamefully, syphilis could eat away at the cartilage and cause a facial deformity known as 'saddle nose', marking out the sufferer for ridicule and censure. In Sarah Cowper's diaries, she remembered hearing gossip about an acquaintance whose nose had collapsed due to the 'foul disease', which she had caught from her wayward husband. Cowper's informant reported that despite her misfortune, the lady was 'content', and 'airy, brisk, and a great dancer'. Cowper, with her typical acidity, wrote:

* In London there were hundreds of 'bawdy houses' and thousands of sex workers, many congregating on the aptly named 'Cokkes Lane', 'Petticoat Lane' and 'Gropecunte Lane'.

'methought that was more than enough, for by no means should any woman dance without a nose, tho' never so innocently lost.'[17]

If a person suspected they had become infected with the dreaded pox, the gold-standard remedy was to purge the poison with mercury, administered by an apothecary. After taking a mercury-containing medicine, the patient would then be bundled up in blankets and placed in front of a roaring fire, with the room's windows and doors shut. The effects were spectacular. Wrapped up like a giant, poxy baby, the swaddled patient would sweat buckets and salivate profusely, to the extent that if the treatment was often repeated, they could lose all their teeth. As the physician William Salmon assured his patients, 'you may know when the salivation is in beginning, by the…vexation of mind and body, diminution or entire loss of appetite, heat of the mouth, swelling and soreness of the gums and lips, thickness of the tongue, blisters in the cheeks, stinking of the breath, whiteness of the palate, pain of the teeth, and indisposedness to sleep'.[18] Such a 'cure' may seem barbaric, but because of the nature of syphilis – worsening, then spontaneously remitting – it appeared to work, and thus was widely used for several hundred years despite its ruinous consequences.[19] It was therefore not lost on the medicine-buying public that this was also an effective, and particularly cruel, murder weapon. Like the court, the medical marketplace could be a dangerous place, and the apothecaries controlled its most deadly products.

The campaigner: Nicholas Culpeper

Despite the scandals, the professional rivalries and the ever-present risk of killing a patient, the apothecary trade flourished. The main reason was that there were simply more people around to buy medicines. The population of England and Wales doubled in the sixteenth and seventeenth centuries, from around 2.3 million in 1520 to about 5.5 million in 1700. Most of that growth took place between 1520 and 1600; during the seventeenth century, famine, plague and war at home and abroad kept birth rates (and life expectancy) down. But the population of London exploded over the same period, from about 200,000 in 1600 to over 550,000 in 1700. It was little wonder that the capital's residents sought out new practitioners, and new means of accessing healthcare. Though their official remit was to make the medicines prescribed by physicians, most apothecaries supplemented their income by giving medical advice and making house calls for their sick patients. But what if medicine could be taken out of the hands of physicians and made accessible to those who needed it most – the poor? What if, with a little education, each man or woman could become their own physician? This was the premise which drove one enterprising young apothecary to author a book of Renaissance medicine so influential that it has remained in print for over 360 years: *Culpeper's Herbal.*

Nicholas Culpeper's eventful life was arguably written in the stars. According to the astrologer John Gadbury, at the moment of his birth on 18 October 1616, the complex interaction of ascendant heavenly signs, houses and latitudes over Ockley in Surrey pointed to 'brevity', with a temperament inclined to the choleric and melancholy.[20] Mercury, the sign of ingenuity, was the strongest planet in the infant's horoscope, but it was mingled with others which spoke of a tendency to lose money as quickly as it was gained. He would have numerous enemies from the fields of medicine and religion, many children and few faithful friends.

The fiery planet Mars in the eighth house was often an indicator of violent death, but in this case mingled with other planetary influences and 'seems rather to portend…to die of a consumption'.[21] It would have been a startlingly good prognostication if it hadn't been for the fact that when Gadbury was writing it, Culpeper was already dead.

The fact that this thirteen-page 'nativity', or birth horoscope, was included in the posthumous *Culpeper's School of Physick* alongside a biography of Culpeper and a letter from his widow shows just how famous he had become. Like his contemporary Hannah Woolley, Culpeper's writings made him a minor celebrity, and the public craved details of his personality as well as his medical knowledge. Nonetheless, there was undeniably something star-crossed about this life, which, though short, contained enough twists and turns for several lifetimes.

Culpeper's father had died just before he was born, so he was raised by his mother, Mary, in his grandfather's household. A dour Puritan minister, William Attersole put his grandson through a rigorous educational programme equipping him with Latin, ancient Greek and detailed theological knowledge, and it was naturally assumed that after a spell at Cambridge University, the young Culpeper would dutifully follow his patriarch into the ministry. But in 1634, aged eighteen, he fell in love.[22] The young scholar and his unknown mistress planned a runaway marriage, heading from Cambridge to the south-coast town of Lewes where they would stay until their parents came around to the idea. On her way to the wedding, however, the girl was struck by lightning and killed.

Culpeper, who had long been interested in astrology and occultism, took the freakish tragedy as a sign. Despite his mother's pleas, he flatly refused to return to Cambridge, which was just as well because his grandfather, infuriated by Culpeper's rebelliousness, refused to fund any further studies. Instead, he paid £50 to bind his wayward grandson as apprentice to a member of the Society of Apothecaries and washed his hands of him. Even this was a concession made reluctantly, and when Attersole died in 1640, he reiterated his ire by leaving Culpeper a measly forty shillings, while another grandchild received £400. Culpeper didn't much care. By that point, he was busy courting the 'two mistresses' of medicine and matrimony. After his first master went bust and fled to Ireland, Culpeper was placed with another London apothecary, Francis Drake (no relation to Elizabeth I's favourite privateer), and completed his apprenticeship alongside an older trainee named Samuel Leadbetter. Unlike Culpeper, Leadbetter had the money to buy himself into the Society of Apothecaries when he finished his apprenticeship, and the two friends worked together in a shop in Threadneedle Street near Cheapside. Culpeper had also found new love, in the form of a fifteen-year-old merchant's daughter named Alice Fields. With cash from Alice's wealthy father, they set up home in Red Lion Street in Spitalfields, an unfashionable but affordable area outside the city walls.* Though he may not have known it, Culpeper was no more than a few minutes' walk from many thousands of bodies buried around the old St Mary's priory and hospital; victims of numerous disasters, but especially the Black Death of 1348–53.**

* This roughly corresponds to what is now Commercial Street, which runs along the east side of Spitalfields Market (established later in the seventeenth century).

** St Mary's was an Augustinian priory and a 180-bed hospital for the poor (run by the monks and nuns of the order) from about 1197 until it was dissolved by Henry VIII and his agents in 1539.

It was an exciting time to be a young man in England; changes were afoot which would reshape the nation forever. For as long as there had been kings and queens, there had been rumblings about those monarchs and their claims to power. At times, such as when Elizabeth I died without an heir to the throne, or when Henry VIII started changing wives as often as codpieces, those rumblings became tremors. But in the years leading up to 1642, the tremors became an earthquake. Charles I had inherited debts from Elizabeth I and James I, who had both borrowed to fund overseas military action. Though he made peace with France and Spain, he was still knee-deep in the expensive project of colonising Ireland, and had tried everything to raise funds (his better schemes included appropriating £130,000 of silver bullion from the Tower of London Mint and utilising a long-forgotten piece of law to fine every knight of the realm who had failed to turn up to his coronation). What was needed was Parliament's permission to raise existing taxes and levy new ones, but not only were they unprepared to give it, they were openly hostile to Charles. After an ill-conceived attempt to arrest five members of Parliament for high treason, the last thread of civility finally snapped, and open warfare between Charles's 'Cavaliers' and Parliament's 'Roundheads' began in 1642.

Culpeper was in on the fray, and firmly on the Parliamentarian side. Partly, this was a religious conviction. Despite rejecting a career in the Church, the young apothecary was a committed Puritan, and he, like many others, resented Charles's attempt to impose uniformity on religious worship. Several prominent Puritans, including William Prynne, Henry Burton and John Bastwick, had lately had their ears cut off and their cheeks branded with the letters S and L ('seditious libeller') as punishment for publishing tracts which criticised Charles and the Archbishop of Canterbury, the much-hated William Laud. Joining the military also offered a convenient escape, since things in London were becoming uncomfortably hot. Shortly after his marriage, Culpeper had fought in a duel – over what remains unknown. Fortunately for both parties, he had not killed his opponent, but he nonetheless had to flee to France for three months to evade arrest.[23] Then in 1642, he was accused of witchcraft by a patient, Sarah Lynge, who claimed that she began to waste away after consulting Culpeper.[24] He was acquitted in January 1643, but he remained vulnerable, in large part because of his involvement with astrology.

Many medical practitioners used the movements of the heavens to discern the root cause of an illness. In fact, they considered it due diligence. Forman, whom Frances Carr had consulted, likened medicine without astrology to the case of a man who tied a string tightly around his arm, covered it with his sleeve, and went to the doctor. The incurious doctor might prescribe a dozen medicines, but without knowing the true cause, none of them would work.[25] On a similar basis, Culpeper saw much of his own work as 'astrologo-physical' and himself as 'student in physick and astrology'.[26] The issue was that he did not confine himself to looking up the planetary sources of people's headaches. He also made astrological predictions that were clearly a product of his anti-monarchism, as were those of fellow prognosticators such as the famous William Lilly.[27] However, the stars did not align for his military career. Culpeper joined the Parliamentarian forces in August 1643, full of righteous zeal. Just over a month later on 20 September, he was shot in the shoulder at the Battle of Newbury. He returned to his day job scarcely older, a little wiser, but with a new understanding of what it felt like to stare death in the face.

Back home in London, life carried on. Culpeper, now set up on his own, was seeing forty patients in a morning. He was a skilled apothecary, but what marked him out from others of his profession was his university education, fashionable long hair and unshakeable self-confidence. Two of those attributes led him into the next and most important stage of his career. In 1617, the College of Physicians had published *Pharmacopoiea Londinensis*, or 'The London Dispensatory', the definitive guide to medicinal ingredients and formulas. They quite deliberately published this work in Latin, a language familiar to the physicians but not many apothecaries. Most of the book's intended users would have been able to pick out descriptions of the various substances described in the *Pharmacopoiea*, but not follow the detailed and often verbose explanations of their proper preparation. Instead, they relied on the craft knowledge passed down from master to apprentice, largely successfully.

The College of Physicians spent much time and energy trying to prove that apothecaries had been preparing their prescriptions carelessly, but the apothecaries countered that they were having to correct requests sent by physicians for medicines containing wildly inappropriate quantities of dangerous substances. For example, in 1633, the Society of Apothecaries revealed that Richard Glover, the apothecary at St Bartholomew's Hospital, had been asked by the hospital's physician William Harvey to provide a medicine containing sixty grains of colocynth (a Mediterranean vine yielding bitter fruit) rather than the usual five.* Because Harvey was the most eminent physician in the country at that time, Glover complied,

* Despite this failure, Harvey was a gifted experimenter who was later responsible for detailing the circulation of blood around the body.

but the patient promptly died as a result.[28] In this professional tit-for-tat, Culpeper had an advantage. The harsh tutoring he had received from his grandfather meant that his Latin was as good as anybody's. Together with his friend, fellow Puritan and publisher Peter Cole, he saw an opportunity – translate the *Pharmacopoiea*, become famous, make lots of money and enrage the physicians all in one glorious swoop.

Culpeper's translation came out in 1649, the same year in which Oliver Cromwell and his supporters shocked the world by cutting off the head of the king. Culpeper's treatment of the *Pharmacopoiea* was less violent but just as decisive. He faithfully translated the physicians' 1617 list of ingredients and their properties from Latin, from borage and turmeric to the brains of sparrows and the yard (penis) of a stag. At the end of the section, however, he added a note that said he'd only included it because the publisher ordered him to do so. The rest of the book consisted of a translation of the college's recently updated list of ingredients, with the vital addition of Culpeper's forthright tips on how and when to use each one. In general, Culpeper argued, physicians relied too much on novel foreign substances and complicated recipes and not enough on local, basic cures, for 'my opinion is, that those herbs, roots, plants &c which grow near a man are far better and more congruous to his nature than any outlandish rubbish whatsoever'.[29] He levelled even sharper criticism at their medical compound recipes. The college's recipe for poppy water, for instance, instructed the maker to 'Take of red poppies, four pound; sprinkle them with white wine two pound; then distil them in a common still; let the distilled water be poured upon fresh flowers and repeated three times; to which distilled water add two nutmegs sliced; red poppy flowers a pugil [large pinch], sugar two ounces; set it in the sun to give it a pleasing sharpness.' Culpeper faithfully relayed the recipe, then added his own (accurate) assessment: 'then it will be a pretty water good for nothing'.[30]

The new dispensatory published by Culpeper staked a bold claim for the apothecaries' expertise and asserted their right to practise without irksome interference. By combining the newest list of ingredients and compounds with additional information straight from an experienced medicine-maker, it also killed the commercial chances of the newly republished *Pharmacopoiea*. In fact, it empowered patients to choose their own remedies with hardly any need for medical intervention at all (for this reason, the Society of Apothecaries came to resent the book, and Culpeper). There was very little in the way of public health provision at this time apart from the hospitals, which had been overflowing even before the war. People were made destitute by medical expenses, but Culpeper told anyone who would listen that much healthcare could be got for free, simply by amending one's diet and taking simple herbal syrups and waters. Yes, it was true that 'there is great danger in physick', and '[m]y intent in publishing books of physick in English, is not to make fools physicians'.[31] Like Woolley, he was wary of ordinary people meddling with potent medical ingredients. But, he insisted, 'It is a base dishonourable unworthy part of the College of Physicians of London to train up the people in such ignorance that they should not be able to know what the herbs in their garden are good for.'[32] That theory would soon be tested in Culpeper's most famous work, *The English Physitian*, which became known simply as *Culpeper's Herbal*.

Compared to other field-defining texts like Vesalius's *Fabrica*, *Culpeper's Herbal* was a modest little book. Its first edition, published in 1652, is ninety pages long, just about small enough to fit in a coat pocket and unadorned by any elaborate frontispiece. It cost only three pence, within the reach of all but the poorest readers. Like all Culpeper's books, this one was sold by Peter Cole in his shops in Cornhill near the Royal Exchange, where it was the jewel in the crown of his portfolio of popular medical works. The ire of the College of Physicians excited rather than intimidated Cole, who was like Culpeper a natural iconoclast.[33] The *Herbal* promised its readers that it would show them how to keep themselves healthy using medicines that cost no more than a few pence to make, and plants that could be found on English soil. Apples, for instance, could be eaten to drive out worms, their leaves boiled for a drink to help agues (fevers), or, when rotten, applied to eyes 'blood-shotten, or enflamed with heat', while the distilled water of fresh apples helped 'procure mirth, and expel melancholy'.[34] Unripe blackberries were good for diarrhoea while the ripe fruit and flowers were both a 'powerful remedy' against both snake bites and, applied externally, piles.[35] Chickweed, which grew plentifully along roadsides and on untended ground, could be juiced or made into an ointment with hog's-grease, and was helpful for everything from swellings and redness of the face to 'cramps, convulsions and palsies', redness in the eyes, earache and ulcers.[36] Cramped sinews were to be dressed with a poultice made from chickweed boiled in wine with rose-leaves then cooked with sheep's feet. Don't know how to make a poultice? A handy guide appears near the back of the book, along with instructions for making syrups, pills, juleps (sweetened drinks) and every other form of commonly used medicine. It was in many ways a commercialised version of the recipe books used by domestic healers, imparting knowledge which had been lost by the dislocation of city-dwellers from their families.

Culpeper's Herbal pulled back the curtain on the mysterious art of medicine, and in doing so became a publishing phenomenon. It went through two editions in the same year it was printed, fifteen before 1700, and over a hundred to the present day, making it one of the most popular medical texts of all time (for comparison, *Gray's Anatomy* has gone through a measly forty-two editions since 1858). Culpeper, however, would not live long to enjoy the fruits of his labour. His old shoulder wound bothered him, and he was weighed down by the appetite of the public – and Cole – for more and more texts. Between 1642 and 1653, he authored or translated thirty-eight works, proclaiming that he meant to 'wear out' his sickly body 'for my country's good'.[37] By the tail end of 1653, he was suffering with a consumption – likely tuberculosis, worsened by his lifelong tobacco habit – which left him a mere 'skeleton, or anatomy'.[38] He died at home in Spitalfields on 10 January 1654, leaving behind his twenty-nine-year-old wife and the one daughter who had survived out of his seven children.

Alice Culpeper, ever practical, found ways to monetise her late husband's legacy. Nicholas had left her many unpublished manuscripts to be printed and sold by Cole – at one point, she claimed to have seventy-nine manuscripts in her possession in addition to seventeen left with Cole (though in the end only twenty-two actually saw the light of day).[39] Shortly after burying her husband, she started selling a popular medicine known as *aurum potabile*, 'drinkable gold'. In life, Culpeper had scoffed at this preparation as one of the physicians' expensive but useless remedies, but people still bought into his brand, enticed by the steady stream of titles: *Culpeper's Last Legacy* (1655), *Mr. Culpeper's Ghost* (1656) and, handily, *Mr. Culpeper's Treatise of Aurum Potabile* (1657, almost certainly authored by Alice's second husband, the astrologer John Heydon). Among those who visited Alice in Spitalfields was none other than Reverend John Ward, who found her 'a very ingenious woman'.[40] When sales of the drinkable gold started to dry up, Alice applied for a licence to practice midwifery, maybe hoping to redress her own sorrowful history of child loss. Perhaps she had in mind the advice of *Mr. Culpeper's Ghost*; that is, the essay written in his guise by his old friend Cole. In this text – part advert, part prophecy – 'Culpeper' advises his readers to sustain the democracy of medicine against the malign influences of the physicians and monarchists. 'Buy these [medical] books while you can get them, study them well, and keep them warily when you have them, and by this means you will cause more to be published in this kind.'[41] The 'Kingly tyranny', he worried, might any day be restored, and then books such as his would be banned. 'YOU KNOW NOT WHAT TIMES ARE COMING', he warned. In many ways, this last, posthumous prognostication would be proved right.[42]

* * *

On 4 December 1665, Culpeper's publisher Peter Cole hanged himself from the rafters of his Leadenhall warehouse. It is unclear why he chose to end his life, especially given the harsh penalties doled out for suicides at that time. Suicide was deemed 'self-murder', a sin in the eyes of both God and the state. If the victim-perpetrator was not one of the two percent deemed '*non compos mentis*', then their estate would be forfeit to the Crown, an institution Cole despised. Cole may have been disheartened by the restoration of Charles II to the throne in 1660, an event which brought with it a spree of book-banning and book-burning. As a known anti-monarchist and Puritan, Cole was probably at the sharp end of the wedge. Nonetheless, he was not destitute. His will – which went unacted due to his suicide – bequeathed land in Suffolk to his brother's son James, 'leases, goods and chattel' to his brother's other children, and a combined £510 to friends and colleagues.[43] He had the document drawn up less than three weeks before his death, suggesting that whatever made him want to die had recently been playing on his mind. Maybe the increasingly hostile environment of the city he called home had worn him down, or perhaps he had fallen victim to some lingering sickness like the consumption that had claimed his friend's life. There is another possibility. Cole knew well the terrible impact of the plague, having lately printed a bill of mortality which documented lives lost during the outbreaks of 1593, 1603, 1625 and 1636. When he died, the plague was sweeping London with more ferocity than ever before. Is it possible that Cole had his will drawn up because he believed he had been in contact with an infected person, and then killed himself when the first symptoms struck?

If so, Cole would have been trying to avoid a fate that had claimed many of his best authors and oldest friends. As we have seen, when the plague struck London in 1665–6, most of the physicians followed their wealthy clients to the countryside. By and large, the apothecaries stayed put. Some of them certainly had the resources to flee; Richard Ingolsbee had a house in Westerham, Kent, and William Firmyn owned property in Hampshire, while several others left estates running into thousands of pounds.[44]

What motivated them to stay was probably a mixture of conscience and commerce. With most of the physicians and many of the surgeons gone, apothecaries were one of the largest providers of medicine to the sick and fearful urban population. Much healthcare was still provided by women and unlicensed healers of various kinds, but even these people needed a source for certain medicinal ingredients. In several of London's hospitals, apothecaries effectively deputised for physicians who had left their posts. At Bethlem Hospital, the trustworthy James James (the most unimaginatively named man in England?) was largely in charge, while at St Thomas's, the queen's apothecary and ex-Royalist soldier William Rosewell was steering the ship. In St Bartholomew's Hospital, Francis Bernard took over so adeptly that he was later granted fellowship of the College of Physicians. The apothecaries' steadfastness did much to boost their popularity in the years after the plague.

But there were also more altruistic motives at work. Apothecaries were deeply rooted within their local communities, as was demonstrated in William Boghurst's *Loimographia, an Account of the Great Plague of London*. Based in St Giles-in-the-Fields, Boghurst, like George Thomson three miles away, decided early on in the epidemic that he 'must not be nice [fussy] and fearful' when it came to dealing with plague patients. Instead:

> I commonly dressed forty sores in a day, held their pulse sweating in the bed…held them up in their beds to keep them from strangling and choking…commonly suffered their breathing in my face several times when they were dying, ate and drank with them…sat down by their bedsides and upon their beds discoursing with them an

> hour together if I had time, and stayed by them to see the manner of their death, and closed up their mouth and eyes…then if people had nobody to help them, (for help was scarce at such a time and place) I helped to lay them forth out of the bed and afterwards into the coffin, and last of all accompanying them to the grave.[45]

The price the London apothecaries paid for their steadfastness was profound. The historian Thomas Whittet estimates that about 1,100 apothecaries and their apprentices were in London when the plague struck, and most of them stayed in the city. The records show that at least five percent of them died. But judging by those whose names were mentioned just before the plague and do not appear again afterwards, the real death toll may have been closer to forty percent.[46] In just over fifty years, practitioners of this craft had been feared as criminals, been ridiculed as incompetent and lost nearly half their colleagues. But they had also formalised into a society, produced a publishing sensation and fought the physicians to secure their place in the medical marketplace.

The Birth of the Virgin. Mary's mother Anna receives refreshments from attendants while the midwife gives Mary her first bath. Woodcut after A. Dürer, 1503. Wellcome Collection.

CHAPTER FIVE

In the Beginning: Midwives

Everybody agreed that Elizabeth Freke was nearly dead. Her husband Percy, Aunt Freke, Lady Thinn and Lady Norton had listened to her fading screams as she lay in labour for almost five days, unable to deliver her first child. In these circumstances, there seemed no way that the baby could have survived, and Elizabeth later recalled that 'all my four midwifes affirmed him several hours dead in me'. There was only one thing for it. The surgeon was called, and he explained to Percy that the infant had to be cut up in the womb and 'taken in pieces'. Reluctantly, Percy agreed. But while the surgeon was 'putting on his butcher's habit', another midwife appeared, asking to be given a chance. She'd been sent by a friend of the family, who swore she could work miracles. Elizabeth thought she'd been sent by God.

For 'three hours in her shift' the midwife worked, and at the end of it, a baby boy was dragged, grey and silent, into the world. He appeared to be dead, for he had 'several great holes in his head' from the hooked implements with which they had tried to prise him out. But somehow, he was breathing. 'God raised him up by life' enough to be baptised 'Ralph' after Elizabeth's beloved father. A skilled surgeon tended his wounds for a month while Elizabeth lay 'so weak as not able to stir'. But it was the anonymous midwife who had given baby Ralph a chance.[1]

* * *

On average, a Renaissance woman would give birth five times in her life.[2] It was *supposed* to be a joyous occasion – after all, the Church taught that procreation was the holy purpose of marriage and the highest calling of a woman's existence. But as Elizabeth Freke found out, it was also fraught with danger. The rate of maternal death in this period was around one percent, one hundred times higher than today.[3] There was a raft of factors, from undiagnosed and untreatable eclampsia to puerperal infections and post-partum blood loss. Behind many of these tragedies, however, was the bare fact that when things started to go wrong during childbirth, options for fixing them were profoundly limited. Surgeons would not countenance caesarean sections on living mothers, since opening the abdomen was virtually a death sentence.* That left two routes: cutting a dead child into pieces in the womb or trusting the manipulations of the midwife.

Renaissance midwives are an odd case among the many medical practitioners of this period. They were less regulated than some other kinds of medics, and their training less formal, yet they were arguably the most important source of medical care for at least half the population. They were often highly skilled, with years of experience, but their practice was irregular. Some midwives attended births every week, others only a few times a year. What is more, their role entailed as much community service as medical intervention. In many places, the midwife – whose title literally meant 'with-woman' – was the person most familiar with both female anatomy and local gossip. Midwives might be

* In *Macbeth*, the eponymous antihero is killed by Macduff, who reveals that he was born by caesarean section, and therefore circumvents the witches' prophecy that Macbeth cannot be killed by anyone 'of woman born'.

called upon to inspect women suspected of procuring abortions or smothering their infants at birth, to search alleged witches for signs of an extra nipple with which they had suckled the devil and to give evidence in cases of neighbourly disputes or domestic violence. When Frances Howard (later Carr) sought to annul her disastrous marriage to the Earl of Essex in 1613, it was midwives who were called to attest that she remained a virgin. Yet despite their importance, they are a spectral presence in the historical record, moving through the most pivotal moments of people's lives like ghosts, seeing and hearing everything but slipping by unnoticed. Unidentified midwives appear everywhere – putting flesh on the bones of their stories is a more difficult task.

* * *

Part of the reason midwives can be hard to trace is that so much about their role was fluid. Midwives were licensed by the Church of England throughout the seventeenth century, with the exception of the interregnum years. Exactly when the Church acquired this licensing power is unclear, but it probably started during the reign of Henry VIII, becoming steadily more important as the urban population grew. Midwives were required to swear an oath which varied from one diocese to the next but included a few key requirements: to provide care to whoever needed it; not to use sorcery; to report all births out of wedlock to the Church; to inform the Church when an emergency baptism had been performed; and not to switch infants or help conceal births. The last rule shows how concerned society was with keeping tabs on family ties. In a system where wealth and titles passed from fathers to sons, women's chastity was an economic concern as well as a moral one, and authorities were particularly anxious to ensure that women did not pass abortions off as miscarriages or infanticides as stillbirths.[4] As such, the testimonials each midwife was required to produce for her licensing had to emphasise that she was not only a skilled deliverer of babies, but an upstanding member of the community. In February 1693, for instance, six women from Herefordshire assured their local bishop that 'Frances Thomas the wife of Edward Thomas of the parish of Stretton in le Dale…is a woman that is very fit and skilful to take upon her the calling and office of a midwife and…is ready to assist and help her neighbours in their distress and to go to the poor as well as the rich.'[5]

However, not all midwives wanted to jump through such hoops, and just because a midwife was unregistered did not mean she was inferior. After all, the process of registration was tiresome and costly, and unlike the upwardly mobile physicians, midwives seldom grew rich in their trade. Parish records show them being paid between two shillings sixpence and ten shillings for delivery of poor women, though fees for attending gentlewomen were likely much higher.[6] The Swiss physician Felix Wurtz warned tight-fisted English customers that midwives and nurses needed to be paid enough that they didn't have to take on 'rough works, whereby their hands are made hard'. After all, he argued, 'if hands are kept clean, because their work in hand is about silk, fine linen, laces of gold or silver, is man not more precious and worthier to be kept clean than all these?'[7] Physicians and surgeons often branded midwives 'ignorant' and 'conceited', but the best of them had years of experience, gained not in dusty lecture halls or macabre anatomy theatres, but at the elbow of other midwives, often close kin. Ann Martin, of St Martin's-in-the-Fields in

London, mentioned in her 1678 licensing testimonial that she was following in the footsteps of her mother, who'd been licensed in 1611 (and had probably brought some of Ann's clients into the world).[8] Almost all of these 'traditional' practitioners were women, though as we will see, they were soon to face competition from 'man-midwives'.

Given that midwives learned by experience and said so little about studying the theory of their craft, it might seem strange that there was a healthy trade in books about midwifery in the seventeenth century. Nicholas Culpeper wrote a volume on midwifery called *A Directory for Midwives* in 1651, while other authors offered their own variations on the theme, from *The Compleat Midwife's Practice* (Thomas Chamberlayne, 1659) to *The Ladies Companion* (Edward Thomas, 1671), or *Every Woman Her Own Midwife* (Anon, 1675). The most influential text, however, was written by an author with the double disadvantage of being both female and a total unknown.

Jane Sharp's *The Midwives' Book* was published in 1671 by the London bookseller Simon Miller, the first book on midwifery by a woman. In content, it reproduced much of Culpeper's work, but its appeal came from Sharp's distinctive authorial tone: forthright, detailed and unapologetic. Sharp told her female readers not to feel abashed about sex, pregnancy and childbirth, for 'we women have no more cause to be angry, or be ashamed of what nature hath given us than men have, we cannot be without ours no more than they can want theirs'.[9] She emphasised the importance of female sexual pleasure, providing details of 'the clitoris [which]...makes women lustful and take delight in copulation'.[10] To back her claims, Sharp emphasised her thirty years of midwifery experience. Apart from this, however, her life remains an enigma. We do not know where or when she was born, whether she was married or had children of her own, or where she lived when she wrote the book. Given that it was dedicated to Eleanor Talbot, the elderly sister of the Earl of Shrewsbury, Sharp might have had connections in the Midlands (the Shrewsbury family seat, Grafton Hall, was located near Worcester), but even this is guesswork. Whatever the truth, her credentials seemingly convinced the book-buying public, for the text was reissued in four editions by 1725 – not phenomenal compared to blockbusters like Hannah Woolley's *Queen-like Closet* or Culpeper's *Herbal*, but certainly good going for a book on this subject, especially one which ran to over four hundred pages.

Who was buying this doorstop of a book? Probably, some midwives picked it up as a reference text, but it was by no means a given that all such women could read – the overall literacy rate for women in the period hovered at just twenty-five percent. Moreover, midwives came from many strata of society, from the upwardly mobile mercantile classes to families on the breadline. When Mary Presse and Sarah Morris testified to the skills of prospective registered midwife Martha Herring in 1614, neither could sign their name to the document prepared for them by a scribe (Sarah made a shaky 'SA' while Mary simply put an X). It is likely, then, that *The Midwives' Book* was being purchased by young people of both sexes eager to learn about the birds and bees. Given the flowery explanations of reproductive anatomy found in many contemporary books, it was little wonder that even well-educated people could labour under dangerous misapprehensions about sex and pregnancy. For all his medical reading, John Ward recorded in his diary that he understood the foetus to breathe 'probably by air received from the woman's privities'.[11]

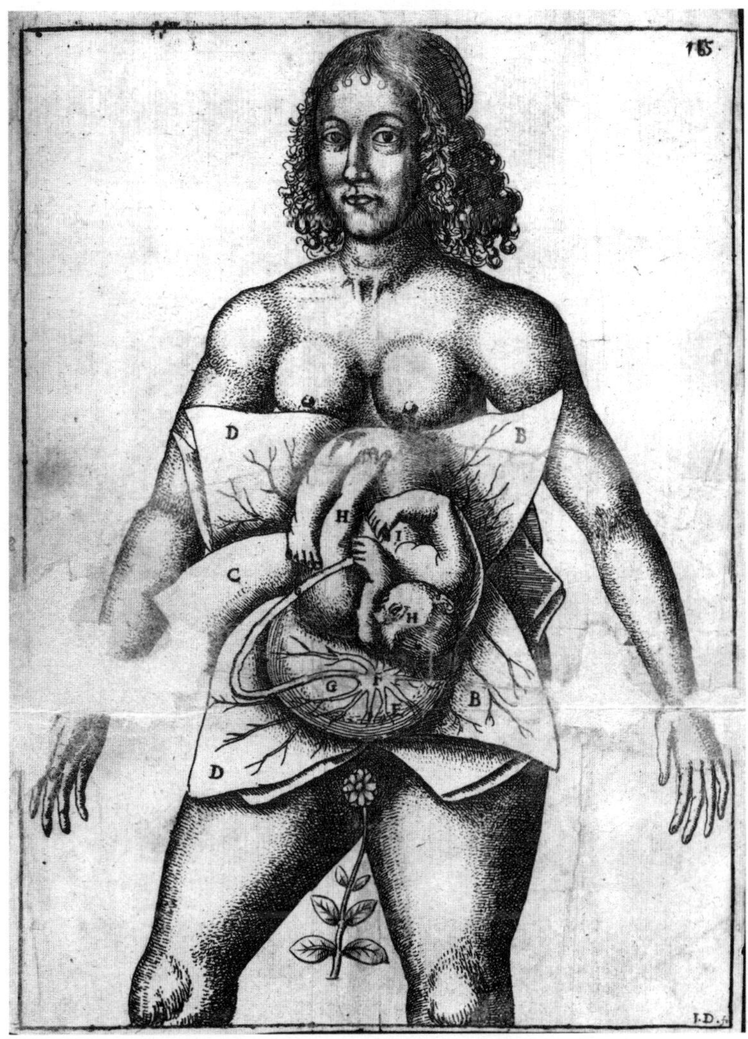

Illustration of the foetus inside the body, from Jane Sharp's *The Midwives' Book*. The Wellcome Collection.

Added to this was the strong belief that the female body was a mysterious and menacing thing. Some medical practitioners held that male and female anatomies were versions of the same template, with women's reproductive organs merely retained inside their bodies because of their lack of 'perfecting' warmth (by contrast, sperm needed to be stored outside men's abdomens because otherwise it would be spoiled by their natural heat). Sharp explained how 'Galen saith that women have all the parts of generation that men have, but men's are outwardly, women's inwardly…the neck of the womb which is the passage for the yard to enter, resembleth a yard turned inwards, for they are of one length, only they differ like a pipe, and the case for it'.[12] Where the vagina was equivalent to the penis, the ovaries were held to be equivalent to the testicles and the fallopian tubes to the spermatic vessels. The whole system was inverted only because women's bodies were humorally colder than men's: 'so then it is plain, that when the woman conceives, the same members are made in both sexes, but…the parts are either thrust forth by heat, or kept in for want of heat; so a woman is not so perfect as a man, because her heat is weaker, but the man can do nothing without the woman to beget children'.[13]

While their anatomy might have been equivalent in theory, in practice a woman's body did not behave like a man's. Indeed, to listen to (male) physicians, it was a miracle that any women remained alive. Women suffered with menses that came too often, seldom or not at all. They experienced migraines, fainting and breathlessness more often than men and were, at least to the Renaissance eye, far more subject to cancers (the overwhelming majority of cancers in this period were diagnosed in the breasts). Perhaps most disturbingly, they generated 'moles' or 'mooncalves', fleshy masses in the womb that imitated the growth of a foetus but were merely lumps of tissue. This was likely the case in the infamous false pregnancy of Mary I in 1555. At age thirty-eight, Mary took to her bed with what she believed to be her first pregnancy, but after months of waiting to be delivered, her belly went down, much to her grief and shame. Rumours of everything from dropsy to witchcraft flew around the beleaguered queen, but it now seems that Mary may have suffered from a molar pregnancy, a growth which originates in the meeting of egg and sperm but fails to develop into a foetus.

The common factor in the myriad of ills which afflicted women was that troublesome organ, the womb. It was connected to everything else; Sharp wrote that 'the brain, heart, liver, kidneys, bladder, entrails and bones…partake with it: but no part is so much of consent with the womb as the breasts are'.[14] Yet there was more to its power. Medical practitioners of all stripes observed that the womb seemed, in some strange way, to be *sentient.* The physician Thomas Bartholin recalled that 'it is furnished…with a kind of brutish understanding, which makes it rage, if all things go not according to its desire'.[15] When it was discontent, it supposedly moved itself up towards the diaphragm 'which presseth and crusheth up the same', a condition known as 'suffocation of the mother'.[16] In such cases, sweet-smelling items were to be placed under the woman's skirts and foul ones at her nose, to lure and drive the womb back into its proper place before suffocation developed into 'uterine fury', a kind of nymphomaniac madness. Some people even believed that the womb continued to live for a time after the woman was dead.

The midwife, then, operated in a realm which was at once arcane and everyday, natural and supernatural. Her job started well before a baby was on its way. Women and men turned to midwives when they struggled to conceive, and Sharp might advise them to eat the brains of sparrows and pigeons, drink galangal in white wine or take a dram of milk in which boar's testicles had been steeped.[17] Knowing when these measures had actually worked was tricky, since pregnancy testing was still more than three centuries off, but for the wise there were always signs. As well as the giveaway of periods stopping, Sharp advised that women with child might have a weak stomach and 'sowr belchings', swollen and painful breasts, visible veins around their eyes, mood swings and strange cravings to eat everything from dirt to stones.[18] Indeed, she added sagely, 'some women with child have longed to bite off a piece of their husband's buttocks'.[19] If one still wasn't sure, then the woman's urine should be collected and kept for three days 'close stopped in a glass and then strain it through a fine linen cloth'. If the woman was expecting, 'you will find live worms in the cloth. Also a needle laid twenty four hours in her urine, will be full of red spots if she have conceived'.[20]

What midwives knew, but seldom acknowledged, was that being pregnant could be a disaster as well as a joy. When clergyman's wife Alice Thornton discovered she was carrying her ninth child while still recovering from a serious illness in 1667, she admitted that 'if it had been good in the eyes of my God I should much rather…not to have been in this condition'.[21] Abortion was strictly forbidden, prohibited from even being discussed by midwives and doctors. That did not mean it never happened. Tucked away in the pages of the Jerningham family receipt book, safe from prying eyes, was a recipe 'to cause miscarriage', which it was promised 'forces away the birth dead or alive as also the afterbirth'.[22] It contained poisonous sowbread (cyclamen) and birthwort (Dutchman's pipe). Other recipes were more concealed, dressed up as means to 'bring down the courses (menses)' or remove 'false conceptions'. In the recipe book of Elizabeth Freke, a drink was listed 'to prevent miscarrying', but it was not all it seemed. Freke's recipe contained many strong spices such as cloves and cinnamon, which pregnant women were usually counselled to avoid, and she advised that the patient was to drink as much of it as they could bear. It would, she advised, 'strengthen the body and preserve conception if it be true', but 'if not it will bring it away with ease and safety and has done many women good'.[23]

Even for women who wanted to be pregnant, it was a worrying time. François Mauriceau, author of a popular French midwifery text, poetically described how 'Going with child is as it were a rough sea, on which a big-belly'd woman and her infant floats the space of nine months: and labour, which is the only port, is so full of dangerous rocks.'[24] More pragmatically, Charles Lytellton complained that when his wife was pregnant 'she can do nothing but puke'.[25] Judging from the advice in midwifery texts and receipt books, Renaissance women were preoccupied with fears of miscarrying, and justly so.* In her memoirs Freke recalled how in the period 1670–3 'I miscarried twice and had very little of my husband's company, which was no small grief to me'.[26] She miscarried again in 1675, which she put down to her distress at having quarrelled with her mother-in-law 'which went very near me, I never having had one unkind word from my father in all my whole life'.[27] Alice Thornton gave birth to her first child prematurely in 1652 after falling into a fever following a strenuous walk. In a heart-rending passage she wrote how

* This was probably because it was a real danger. Upper-class women in particular consumed quite large quantities of alcohol (between three and nine ounces a day) and there was no advice given to stop this in pregnancy. Many women of all classes also suffered from chronic diseases, while working-class women were frequently malnourished.

> my poor infant within me was greatly forced with violent motions perpetually, till it grew so weak that it had left stirring, and about the 17th of August I found myself in great pains as it were the colic, after which I began to be in travail [labour], and about the next day at night I was delivered of a goodly daughter, who lived not so long as we could get a minister to baptise it…Thus my sweet babe and first child departed this life half an hour after its birth.[28]

Not only was the loss an emotional blow for Alice, it nearly killed her. In the months after her premature labour, she was so ill that 'The hair on my head came off, my nails of my fingers and toes came off, [and] my teeth did shake, and [were] ready to come out and grow black.'[29]

Midwives had a raft of cautions for pregnant women. They were to avoid spiced foods, hard exercise, travel, long baths, excessive bloodletting, purging medicines, strong emotions and even pungent smells. If they feared losing their pregnancy, they could drink a syrup of tansy flowers or one of the unnumerable potions and powders listed in household receipt books 'for a woman in danger to miscarry'.[30] Some physicians advised that women likely to miscarry should have their blood let, though more cautioned against it. Ultimately, though, there was only so much that anybody could do to bring mother and baby safely through the rough waters of pregnancy, and as they approached the port of labour, a happy outcome was far from certain. In the last trimester of her pregnancy, the expectant mother would prepare the baby's linens and, if she was prudent, write her will.[31] Shortly before the main event, she entered the last and most dangerous phase of pregnancy: the confinement.*

* * *

The lying-in chamber was a special place, where women held sway and the busy, noisy outside world was kept at bay. The windows were draped with thick curtains, the room illuminated only by candles. Keyholes were stopped up against draughts and the mother was nourished through her ordeal with broth and a drink of caudle (warm ale or wine mixed with sugar and spices). It was a womblike space, in which a woman could prepare for the journey she was about to undergo with the midwife as her guide.

No men were allowed inside this sanctum; instead, it was populated by the mother, her midwife and her 'gossips', female friends and relatives who gave moral support, helped tend to the new-born and kept the mother entertained during the tedious early stage of labour. The word 'gossip' derived from 'God-sibling', and the gossips were originally those who would later witness the child's baptism. However, gossips were also a practical help. When the clergyman Ralph Josselin's wife Jane gave birth to a son in 1648, her labour progressed so quickly that the midwife could not attend in time, and she delivered with the help of her five gossips. Ralph recorded a similarly frantic incident

* The period of occupying a separate room before giving birth was usually known as confinement, and the forty days of bed rest after the birth as 'lying-in'. However, they were sometimes all simply known as lying-in, and the poorest women – of whom we have very few records – probably didn't get to rest much either side of their labour.

three years earlier in his diary, during the birth of their third child (of ten):

> 24 [November 1645]: I had sought to God for my wife (that was oppressed with fears that she should not do well on this child), that God would order all providences so as we might rejoice in his salvation…About midnight on Monday I rose, called up some neighbours; the night was very light; goodman Potter willing to go for the midwife, and up when I went; the horse out of the pasture, but presently found; the midwife up at Buers, expecting it had been nearer day; the weather indifferent dry; midwife came, all things gotten ready towards day. I called in the women by daylight, almost all came; and about 11 or 12 of the clock my wife was with very sharp pains delivered Nov. 25 of her daughter intended for a Jane…wife and child both well praise be my good and merciful father.[32]

Ralph's main purpose in keeping his diary was spiritual, and his account shows him giving thanks for all the circumstances that had made Jane's birth safe: the cousin was willing to go for the midwife; it was a dry, moonlit night; the horse had escaped but was easily found; the midwife had expected a later birth but was nearby. Ultimately, the stars aligned so both Jane his wife and Jane the baby survived the ordeal. But despite all this, the anxiety Ralph felt for his beloved wife breaks through his carefully controlled narrative.

While the midwife performed vital functions – staunching blood loss from the mother, making sure the placenta or 'secundines' were delivered, cutting the umbilical cord and reviving babies slow to take their first breaths – the birth itself was something which had to happen naturally. It is small wonder that even the most god-fearing women turned to charms and rituals in labour that they would not have dreamed of at any other time. When the Catholic Church had reigned in Britain, monasteries and convents had been full of sacred objects to be lent out to women about to give birth. The Cistercian nuns at Keldholme in Yorkshire had the finger of Saint Stephen, while the Benedictines in Derby had Saint Thomas's shirt. At Sinningthwaite, also in Yorkshire, was the tunic of Saint Bernard, and all over the country were dozens of girdles said to have belonged to various saints or even the Virgin Mary herself. Sir William Clopton of Long Melford in Suffolk claimed to have a splinter of the true cross, which he bequeathed to his son in 1530 with the instruction that it should only be lent out to honest women to assist them in childbirth.[33]

After the Reformation, these items were hidden, confiscated or destroyed, but the hopes and fears which had led women to seek them out did not subside. Instead, the old ways – never very far from the surface – came back to the fore, as folk magic blended with pseudoscience. A favourite among medical authors was an eagle-stone – that is, a hollow geode, sometimes containing another smaller stone, which was said to have 'magnetic virtue' to help draw out the baby when it was held near the woman during labour. Such was its power that Culpeper warned it had to be used with great caution, as 'both child and womb follow it as readily as iron doth the lode-stone [magnet]'.[34] Thomas Lupton's eclectic 1579 book of tips and remedies, titled *A Thousand Notable Things*, suggested that oakfern, sage or mugwort tied to the woman's thigh might have a similar though less powerful effect.[35]

Mostly, however, there was no explanation given for the wide array of objects that could be worn by or placed near a woman to ease her labour, including shed snake-skins, a gourd, a magnet (held in the left hand), a piece of red coral, the bloodied skin of a hare or a horse's hoof.[36]

The one thing that bound all these strange objects and practices together was fear: of pain, of losing a child or of losing one's own life. A baby 'turned wrong' in the womb – a breech birth – was a life-threatening concern. Alice Thornton experienced the panic of a breech birth while in labour with her fifth child in 1657. Living in the remote hamlet of East Newton on the Yorkshire coast, her options for medical help were limited. She remembered how for twenty-four hours she'd felt 'such exquisite torment, as if each limb were divided from the other'. Eventually, the midwife managed to deliver the child, but 'coming into the world with his feet first, caused the child to be strangled in the birth, only living about half an hour'.[37] It was a horribly familiar danger, and Sharp's *The Midwives' Book* furnished its readers with comprehensive diagrams of babies in a variety of difficult poses: feet first; arms outstretched; back first; legs bent. Her advice centred on moving the mother in order to shift the foetus. For instance, if the child 'will come forth with one foot, and the other lifted upward', then she advised that the mother be laid on her back with her legs held up

> that her head be lower than her body; then let the midwife with her hand gently put back the leg that is come forth into the womb again, and bid the labouring woman to stir and move her self, that by her stirring the birth may offer it self the head downward, and if so, you may then

> set her in a chair as she was at first that she may have a natural delivery, but if this cannot be done, then the midwife with her hand must discreetly bring forth that leg that is not yet come forth; but beware she put not the child's hands that lie close down by its sides out of their place.[38]

In her instructions, Sharp showed every bit of her thirty years' experience, but even she admitted that some kinds of presentation – namely the feet-first baby, like Thornton's son – were 'the worst for danger of all the rest', and 'it will be hard to it'. Out of such cases, a new kind of practitioner was born: the man-midwife.

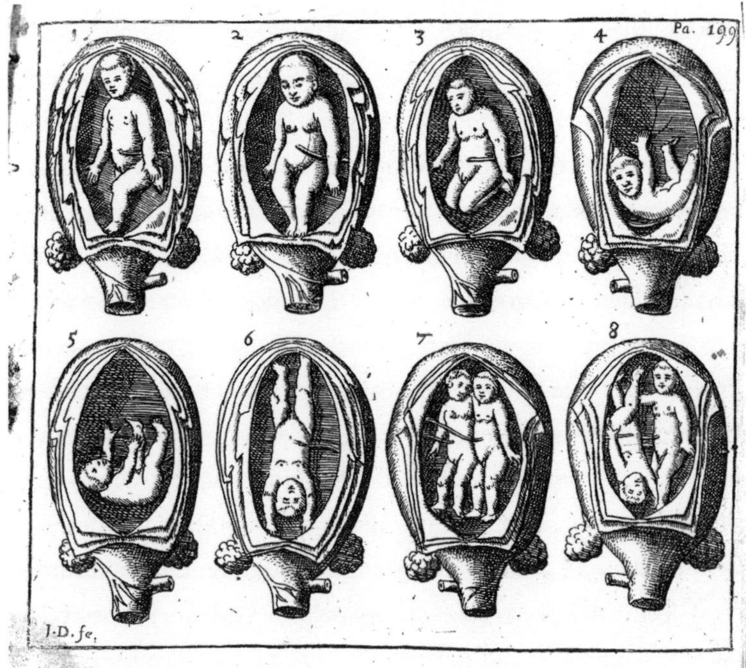

Diagram of babies in different presentations for birth from Jane Sharp's *The Midwives Book*. The Wellcome Collection.

Secret-keepers: the Chamberlens

Renaissance medicine was a family affair. Physicians sent their sons to the universities they had attended, surgeons and apothecaries took nephews and cousins as apprentices, midwives passed down their knowledge to daughters, and receipt books were handed from one generation to the next. Medical practitioners were also family in a larger sense. Frequently at one another's throats, they also wove in and out of each other's lives. Look carefully, and they can be seen in the background of seemingly unconnected scenes, changing the course of events in unexpected ways.

The Chamberlens were a family of medics, connected to the wider medical world in all sorts of ways. In a very roundabout fashion, they arrived in England with the help of Ambroise Paré and Andreas Vesalius, the two men who had tried and failed to save the life of Henry II, King of France, after his jousting accident in 1559. Henry's death precipitated a power struggle between his widow Catherine de' Medici and various French nobles, all played out over the head of his son and heir Francis II. That struggle in turn helped light the fuse on the conflict which had been brewing in France since the 1530s, as 'Huguenot' Protestants butted heads with the majority Catholic population. In 1562, the dispute finally caught fire, and a bloody civil war ensued which would displace 300,000 Huguenots, driving nearly 50,000 to seek shelter in the nearby Protestant stronghold of Britain. Among those weary refugees, landing in Southampton on 3 July 1569, were the barber-surgeon William Chamberlen, his wife Genevieve and their children Peter, Simon and Jane.

Despite their strange accents and rather strict Calvinist religion, the pair, with three young children in tow, would have been welcome. Having lost so many young men in Elizabeth's various foreign wars, England desperately needed the skills and knowledge of these Protestant refugees. In the bustling workshops and warehouses of the city, one might have heard many languages spoken, from guttural high-Dutch to mellifluous Italian. Settled on the south coast, Genevieve went on to have two more children, James and a second Peter (later known as Peter the younger). Between them, the two Peters were to establish a medical dynasty spanning five generations.

Having experienced so much upheaval, the Chamberlens were keen to see their children established in steady trades, and both Peters were trained up as barber-surgeons. They moved to London to ply their trade, where they soon gained a reputation as ambitious upstarts. Both men were censured by the College of Physicians for overstepping their professional boundaries by prescribing medicines. Peter the younger confessed to having treated syphilis by purging in 1600, and his elder brother was a prolific offender against the college's strictures, once ending up in Newgate prison. As we have seen, getting on the wrong side of the college was unfortunate but not difficult, especially since the physicians were at this time trying (and failing) to clamp down on the practice of physic by apothecaries and surgeons. What was more unusual, however, was the interest both Peters took in the conventionally female pursuit of midwifery.

Usually, the presence of a surgeon in the lying-in chamber itself was a signal that something had gone catastrophically amiss and the life of the labouring woman was in danger, as was the case in Elizabeth Freke's pregnancy. The surgeon's job was to extract a dead child (or one presumed dead) if it could not be delivered, and to stitch up the mother if she was badly torn. His primary tool was the dreaded 'crochet', a long blunt hook designed to draw the baby out using a hole made in the skull by the equally grim 'perforator'. For wealthy women, having a surgeon in the house ready to hand at the time of their labour was a reassuring luxury, and in this capacity, Peter the elder attended at the birth of Charles II in 1630 (though the event was managed by a midwife named 'Madame Péronne'). He was, however, a last resort, to be brought in only if all else had failed.

To the two Peters, this seemed a topsy-turvy state of affairs. Surely, they reasoned, it was the physicians and surgeons who ought to instruct the midwives, not the other way around? After all, at that time, every text on childbirth was written by a medical man. The Chamberlens were not alone in thinking this way. In Derbyshire, the physician Percival Willughby had made a name for himself as an expert in obstetrics, focusing specifically on how to manoeuvre babies who presented in the wrong way. There were also a handful of male midwifery practitioners, or *accoucheurs*, operating in the capital, including John Nowell, James Blackbourne, Nicholas Downing and 'Mr. Doughton'.[39] However, the Chamberlens were by far the most vocal. When, in 1616, the midwives of London petitioned to be recognised as an organisation, Peter and Peter eagerly put themselves forward as overseers. Their offer, and the petition, were passed on by privy counsellor Francis Bacon to the College of Physicians, who agreed that midwives were 'ignorant', but swiftly rejected the idea of their becoming incorporated. To press the point, they chided the Chamberlen brothers for 'impudently' implying that they, or any surgeon, could know more about childbirth than physicians.

When the two brothers died (the younger in 1626, the elder in 1631), their efforts to reform midwifery had amounted to little. Nevertheless, Peter the younger left a legacy in the form of eight children, two of whom were physicians. The most influential of these was yet again named Peter, who distinguished himself from his father and uncle by referring to himself as 'Doctor Peter', which handily also highlighted his medical qualifications. Doctor Peter redoubled his family's campaign for regulation of midwifery, but to no avail; the king not only rejected his 1634 petition to set up a Corporation of London Midwives, but cursorily passed it on to the bishops, who insulted Doctor Peter by suggesting that if he so wanted to be a midwife, he ought to apply to them for a licence. It was a bitter blow for a man with a high opinion of his own medical genius, who boasted that he came from an 'Asclepiad-family'.[40] In a text rather ostentatiously titled *A Voice in Rhama*,[41] he complained bitterly that he was being personally victimised by snobbish 'back-biters', and the antipathy of both the College of Physicians and the midwives to his suggestions was evidence of their close-minded attachment to the old ways:

> The objection infers thus much, because there was never any order for instructing, and governing of midwives, therefore there never must be. Because multitudes have perished, therefore they still must perish. Because our fore-fathers have provided no remedy, nor knew any, therefore we must provide none, though we know it.[42]

Doctor Peter was not as persecuted as he imagined, for though he was temporarily expelled from the College of Physicians when his patron Charles I lost his head, he managed to obtain a lucrative monopoly on making baths and bath-stoves in 1649, and in 1661, he was appointed as physician to the newly restored Charles II. Like the other members of his ever-growing family, he was constantly looking for the next big thing, travelling Europe to take out patents on his ideas for wind-powered transport and phonetic writing systems and becoming ever more deeply invested in 'Independent' and unorthodox branches of Protestantism. In 1662 he was obliged to publicly defend himself against allegations of madness, and in 1680, against the even more damning charge of being a Jew. Yet despite the chaos which followed him, he never relinquished his ambitions. Moreover, he was well supplied with prospective helpers. Peter had at least eighteen children with his wife Jane (herself one of sixteen progeny of the Welsh entrepreneur Sir Hugh Myddelton). Of his fourteen sons, four followed in the family business and became physicians, of whom the most prominent was Hugh, born sometime between 1630 and 1634. This meant that in the 1660s and 70s, Peter and his son had the chance to work together, combining the younger man's vigour with the older's experience.* Finally, they could see a change on the horizon.

In 1670, Hugh Chamberlen travelled to Paris, to the Hôtel-Dieu where a young Ambroise Paré had cut his surgical teeth more than a century earlier. Many English medical practitioners studied in France, and Paris was a favourite destination for surgeons wishing to see the latest techniques at work. On this occasion, though, the traveller hoped to gain more than

* The pair stayed in London during the plague of 1665–6, so they probably had some contact with the unorthodox physician George Thomson, as well as the many apothecaries who remained in the city.

knowledge. Once more, the Chamberlen family were on the make, and Hugh had a secret to sell. It was one he valued at ten thousand crowns, a mammoth sum, but he claimed that this secret would help the user to deliver a child in a difficult birth in a matter of minutes – and *alive*.* A delicate negotiation with the French king's physicians ensued (the unpleasantness of Hugh's family having been driven out of the country a century earlier apparently having been forgotten), and in an effort to ascertain whether this device really was worth as much as the visitor claimed, they passed Hugh onto the Parisian surgeon François Mauriceau, with instructions to see what he was about. After a few months, Mauriceau had found what he deemed a suitable test. In the hospital was a 'tiny woman' – perhaps suffering with rickets – who had been in labour for eight agonising days. The baby was head-first, but stuck fast in the woman's narrow pelvis, and Mauriceau himself could do nothing. The scenario was certainly a test for Hugh and his methods, but it was hardly a fair one, since if the woman's pelvis really was deformed by rickets, there was no way the child could have been delivered except by a deadly caesarean section.

Predictably, the attempt failed, though what exactly that meant was never explained. Probably, both mother and child died – a fact conveniently omitted from the medical men's records – and the hoped-for ten thousand crowns slipped away with them. But for Hugh, the visit was not a complete loss, since he secured the job of translating Mauriceau's medical treatise *Traité des maladies des femmes grosses* into English as *The Accomplisht*

* A crown was worth about five shillings, so Chamberlen was asking for the equivalent of about £280,000 in modern terms.

Midwife. In the translator's preface to the text, he took his chance to make the first direct claim in print for his much-whispered-about 'secret':

> my father, brothers, and myself (though none else in Europe that I know) have by God's blessing, and our industry, attained to, and long practised a way to deliver a woman in this case [i.e., the baby stuck head downwards] without any prejudice to her or her infant: though all others...do, and must endanger, if not destroy one or both, by the use of these crochets [hooks]. By this manual operation we can also shorten the time, and lessen the number of pains in a right labour (if there be the least difficulty) without danger, and with advantage to both woman and child.[43]

With the instincts of a born salesman, Hugh pinpointed the fears women had around birth and promised relief. Aware of the public's suspicion of physicians, he emphasised practical knowledge passed down through the family. However, having tantalised the reader with means for safer, shorter, less arduous labour, he immediately followed up with an apology that he couldn't actually tell the reader what this great secret was. The reason, he explained, was that his two brothers and father used the secret, and he could not publish it without injuring their interests. In fact, the Chamberlens made a huge production of protecting their intellectual property in hopes of increasing its value. When they attended births, they had two men carry in a large box covered in carvings, hoping to create the impression of some massive and complicated piece of machinery. They shut all others out of the lying-in chamber and even allegedly rang bells and shook clappers to put would-be copycats off the scent.[44] It was an early and spectacular example of stunt marketing in action.

Hugh and his father expected the titbit in *The Accomplisht Midwife*, combined with their carefully staged medical theatrics, to create such a clamour that they would be inundated with offers, perhaps exceeding the fortune that Hugh had failed to secure from the French king. In the event, things didn't work out quite as they had hoped. The Chamberlens had diluted their brand with their various enterprises. One brother was selling 'anodyne necklaces', cheap beads which he claimed could help ease labour pains. Doctor Peter was still hawking bathtubs. In 1682, Hugh lost his place as Charles II's physician, which he had inherited from his father, because the king wished 'not to admit the said Dr Chamberlain to a place so near his person', and when in 1687 he tried again to obtain a patent vaguely said to 'relate to midwifery', he was unsuccessful.[45] He was in and out of favour with the court and the medical establishment for the rest of his life. Just after his thwarted attempt to gain the patent, he appeared before the College of Physicians on charges of malpractice for having treated the six-months-pregnant wife of John Willmer with bleeding and purging, after which she miscarried. He was conspicuously absent from the birth of James II's son on 10 June 1688, but on 17 April 1692, received one hundred guineas for delivering the future Queen Anne of a son who died almost immediately afterwards. It is not known whether he used his secret in the delivery.

Eventually, the Chamberlen project began to wind down. Hugh had four children, of whom the eldest son – also named

Hugh – carried on in the medical trade and followed his father, grandfather and great-grandfather into man-midwifery. He was, however, a much quieter figure, having seen what ambition and notoriety had failed to do for his forerunners. Where his father and grandfather had often clashed with the authorities, Hugh the younger became a censor of the College of Physicians, imposing their rules rather than breaking them. Meanwhile Hugh the elder, a man of unusually healthy old age, relocated to the continent, where he sold his midwifery secret to the Dutch surgeon Henrik van Roonhuisen. It was not the deal of the century he had long hoped for, but he needed cash. By the 1730s, with all the medicine-practising members of the Chamberlen clan dead, the 'secret' was finally out, and it was better than most people had expected.

At some point in their long dynasty, the Chamberlens had invented and refined a collection of obstetrical instruments, the most important of which was the forceps. This wood and metal device had handles which crossed in the centre like a pair of scissors, with curved spoon-shaped blades for gripping the baby's head. It was a simple tool, but it allowed the Chamberlens to assist in births where the baby was stuck head-first in the birthing canal without using the dreaded crochet, which often fatally perforated the skull of the child and risked tearing the mother internally. In fact, we know exactly what the Chamberlen forceps looked like, because in 1813, several pairs were found underneath the floorboards of Doctor Peter's old home, Woodham Mortimer Hall. With them were a vectis, a sort of long shoehorn for manoeuvring the baby's skull, and a fillet, an arrangement of three thin curved metal strips attached to a central handle, which could surround the baby's head to

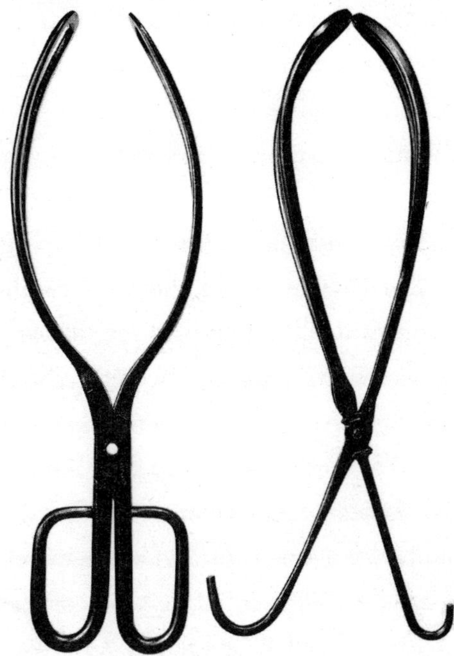

Photograph of the 'Chamberlen forceps' found at Woodham Mortimer Hall, Essex. From facsimiles in the Wellcome Collection. The originals belong to the Royal College of Obstetricians and Gynaecologists.

ease its passage down the birth canal.* They'd been hidden there by Peter's wife Anne – the last keeper of the family treasure.

* * *

If both mother and baby survived the perils of labour, the midwife – male or female – could breathe a sigh of relief. The lying-in

* All three tools were still in use (with some amendments) in the early twentieth century. Chamberlen's forceps design continues in use today.

chamber was, for a moment, a place of joy. For the midwife, however, the respite was short-lived, for there was as much to do after the baby arrived as before.

First, the umbilical cord or 'navel-string' had to be cut. This was not as simple as it might seem. In his book, Culpeper dismissed the popular myth that children in whom the navel-string was left longer would be more fertile in later life (specifically, though Culpeper was too polite to say it, it was believed that boys with umbilical cords cut longer would grow larger penises). However, he subscribed to other superstitions. If the umbilical cord was allowed to touch the ground after it had been cut, the child would be 'subject to an involuntary pissing during its life'.[46] If a piece of the cord was kept and worn by another person touching the skin (he suggested in the inside of a ring), it would protect the wearer from 'falling-sickness' (epilepsy), convulsions and even witchcraft. Culpeper admitted that he had no evidence for this belief, but he had heard it from an honest man, and as he pointed out with straightforward if somewhat flawed logic, 'honest men usually tell the truth'.[47]

The next priority was to ensure that the 'secundines', or placenta, was delivered. Midwives might try to extract the placenta manually, but, as Sharp warned, 'long nails may do mischief'.[48] In 1700, John Clerk recalled the traumatic story of his wife Margaret's death. Early on 20 December, Margaret went into labour and begged John to fetch his father, because 'she was assured that she was dying' (why she wanted John's politician father there is unclear, but she asked him to pray for her). At seven in the morning, she was successfully delivered of a son and it seemed her premonition was unfounded, but 'when everybody had run into my room to carry me the good news, she fell into fainting fits'. A physician and surgeon were summoned who determined that the placenta was stuck to her uterus, and 'this they thought themselves obliged to bring away by force'. Alas, John wrote, 'they were too hasty in their operations, by which she lost a vast deal of blood', and Margaret died in agony at eleven that morning.[49] John could not bear to keep his son, and he was sent to live with an aunt until he was old enough for school.

Margaret's story was harrowing but all too common. Alongside retention of the placenta, the most common causes of death after childbirth were infection and blood loss, and midwives and gossips were responsible for trying to restore the new mother to her full strength as quickly as possible. They would feed her broth and caudle at first, then light meats such as chicken, then spiced wine, bread, butter and sugar. For the first week, her genitals were washed daily with water in which chervil had been boiled, the room kept dark and quiet, and the woman encouraged to sleep.[50] When the baby was clean and tightly swaddled, and the mother out of immediate danger, the midwife's job was done. Nobody knew better than these women that their charges might not survive. Babies born in this period had a one in ten chance of dying in infancy, from a raft of diseases which were hard to identify, never mind cure. Alice Thornton, who gave birth to her first child prematurely and nearly died delivering her fifth, would have nine children during her lifetime. Of those, six would either be stillborn or live less than a year, with babies Christopher, Joyce and William all separately felled by mystery illness which 'struck in many red spots' all over their face and body. Thornton did her best to see these losses as God's will, taking her babies out of a miserable world and into a better one. But this did not mean that she loved her children any less. The autobiography which contained all these accounts was started in 1669 as a response to some local slander, but it became a form of therapy for her losses. Nearly twenty years after the premature birth of her first child, she still remembered each one as if it had been yesterday.

* * *

In 1687, the midwife Elizabeth Cellier wrote to King James II with a proposal entitled 'A Scheme for the Foundation of a Royal Hospital'. It was an audacious plan by any measure. Cellier proposed that the king should give her £5,000 a year to establish a 'corporation' of two thousand midwives, the 'most able and matron-like women', who would attend lectures under the tuition of a man-midwife. The midwives would pay fees for the privilege of belonging to this elite corporation, which would help to fund the other part of Cellier's scheme – a foundling hospital where abandoned or illegitimate children could be raised, take on new names and learn an occupation which would eventually make them useful members of society. Added to this, a dozen lesser houses would be established in the twelve most populous parishes of London 'for the taking in, delivery, and month's maintenance' of unmarried mothers, who would then give up their babies to the care of the institution. The whole scheme, Cellier argued, would help to save thousands of lives; using some rather fast and loose calculations based on the Bills of Mortality, she estimated sixteen thousand women and babies to have perished as a result of stillbirth, death in childbirth, miscarriage or infanticide in the two decades preceding her letter.

It was an extraordinary request. In some ways, Cellier's vision was surprisingly forward-thinking. The infants to be housed in the foundling hospital would be able to choose new surnames for themselves and thus escape the stigma of their bastardy, or 'innocent misfortune' as she delicately termed it.[51] They would remain in the care of the hospital until the age of twenty-one but could stay longer if they wished. Cellier also included in her plans a generous income for herself, 'that nothing may divert her from employing all her industry for the good of those poor exposed children'.

What was most interesting about Cellier's proposal, however, was what she left out. Unmentioned in the text, but surely known to its reader, was that Cellier was the most famous – or infamous – midwife in London. She was a Catholic at a time when this was once again a dangerous faith to profess. The Great Plague of 1665–6 had been blamed on Catholics (as well as Jews, atheists, sinners in general), and when Charles II repealed the harsh laws against Catholicism in 1672, there was a widespread fear that this minority group would somehow take over. In 1678, anti-Catholicism reached a fever pitch following the 'Popish Plot', ostensibly an elaborate plot to assassinate the king, which resulted in the executions of twenty-two Catholics. Shortly thereafter it was discovered that the whole thing had been a Protestant hoax. Against this backdrop, Cellier had been visiting prisoners in Newgate, the same jail which had at various times housed Nicholas Culpeper, Peter Chamberlen and George Thomson. There she met Thomas Dangerfield, a trickster who exploited Cellier's gentry connections before turning on her and accusing her of treason. Cellier was acquitted, but then jailed for libel for alleging (probably truthfully) that the Catholic prisoners in Newgate had been tortured.

One might think that after such treatment, Cellier would want to stay well out of the limelight and far from the gaze of the authorities. At the time of her trials, the popular press had been vicious in their attacks on the 'Popish midwife', denouncing her claims as a 'dunghill of lies and contradictions'.[52] In a response to criticism of her proposal, Cellier claimed that she had been prompted to write because 'though you have often laughed at me, and some doctors have accounted me a mad woman these last four years, for saying Her Majesty was full of children, and that the bath would assist her breeding', she was now vindicated by the queen's pregnancy.[53] James Stuart, later 'the Old Pretender', was born on 10 June 1688, a longed-for living son for the king, who had fathered five stillborn children and a further eight children who died in infancy, in addition to three living daughters. But in that same text, Cellier denounced male-midwives as being about as qualified to teach midwives as a civilian was to 'read a military lecture to Hannibal the Great'. Given the gulf between this scathing critique and her suggestion in the *Proposal* that the corporation of midwives be overseen by a male physician, it's hard *not* to smell a rat. Was the *Proposal* really a newly dressed version of Hugh Chamberlen's recent unspecified and unsuccessful 'patent' application, this time channelled (for a fee) through a well-known female practitioner?

If so, it proved futile. Cellier's petition was apparently granted that November, but the promised midwives' corporation never materialised. In the meantime, Cellier vanished from the historical record. The birth of James's son in 1688 destabilised the kingdom by opening up the possibility of a Catholic heir to the English throne. There was even a rumour put about (possibly by Princess Anne) that the baby was not Queen Mary's and had been smuggled into the birthing chamber in a bedpan. Though the allegation was transparently false, the succession issue so rattled

Protestant nobles that they swiftly executed a coup, replacing the king with his firmly Protestant sister Mary II and her husband William of Orange. Cellier probably left the country before she could be forced out. As for the misfortunate infants for whom she was so concerned, there was a foundling hospital established in London, but not until 1739.

For the next two hundred years, midwifery would be shaped by the handful of strong personalities who had battled over the territory of the female body in the 1600s. It became more medicalised, especially for the aristocracy. The forceps eventually came into common use, for better and worse. They saved lives by delivering stuck babies, but they were always considered a tool for surgeons and physicians, and the reluctance of those men to wash their hands when they went from the morgue to the birthing chamber meant that maternal death rates stayed stubbornly high. Obstetrics was at the vanguard of medicine, a chance for male doctors to stamp their mark. But at heart, midwifery remained a field dominated by women, and the Midwives Act, which made training compulsory for anyone practising as a midwife, was not passed until 1902. In 2017, men made up 0.4% of midwives in the United Kingdom.

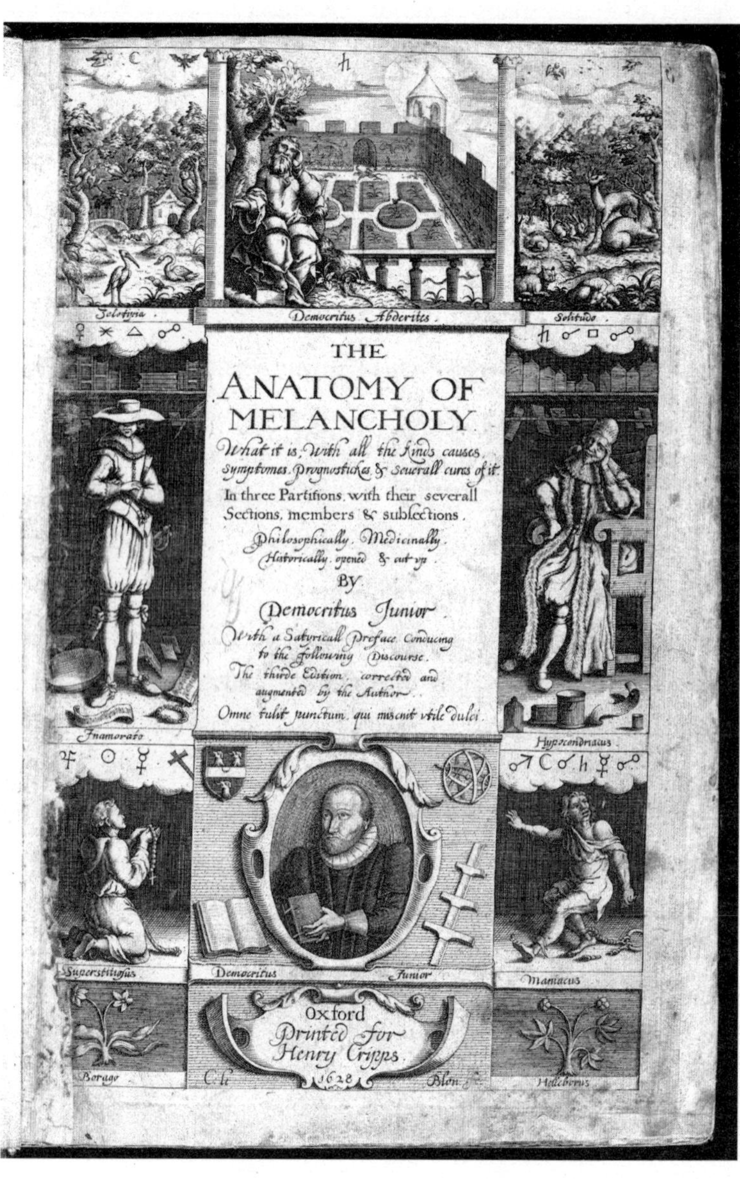

Frontispiece of Robert Burton's *The Anatomy of Melancholy*, 1628. Wellcome Collection.

CHAPTER SIX

Insanity: The Mad and Their Doctors

In a cool, damp chamber on the marshy outskirts of Oxford, a middle-aged scholar in a black gown and small starched ruff is trying to write his way out of depression. It is no easy task; it will be his life's work, and the finished tome will span nearly eight hundred densely printed, meticulously annotated pages. But it is the best way he knows to settle an 'inconstant, unsettled mind', so fascinated by its own sadness.[1]

Recently, the scholar has been researching delusions. In the books that surround him are strange tales: a man who believes he is made of glass, terrified to sit lest he shatter; paupers convinced they are kings and queens; a woman who thinks her belly is full of eels; a host of unfortunates certain they are dogs, chickens, bears, dwarves or giants. One of his favourites describes a young man so sure his nose is of monstrous proportions that he refuses to walk down the street because it is bound to scrape against the buildings on either side.

Today he is engrossed in the story of a madman from Italy, who maintains his bladder holds gargantuan volumes and is afraid to urinate in case he drowns the whole town. The man's physicians have conspired to play a trick on him. They have rung the church bells, and enlisted locals to pretend that there is a fire threatening their home (indeed, there is a small fire, since smoke was needed to add to the illusion – the physicians do nothing by halves). Running to the madman, they entreat him to put out the blaze with his flood of urine, 'whereupon he pissed, and was immediately cured'.[2]

Medicine by deception, a lie which restores the patient to the truth. It is a satisfying end to a tale; but the scholar's discontent is not so easily shifted. It has been his companion for so many years that he is not sure what it might be like to live without it.

* * *

In 1621, this unassuming academic and churchman, Robert Burton, published *The Anatomy of Melancholy*. An immense work of over half a million words, it is a curious text; part self-help book, part scientific treatise, part confessional. For Burton, it was both his life's labour and his therapy; as he confessed, 'I write of melancholy…being busy to avoid melancholy…to ease my mind by writing.'[3] 'Melancholy' in these times meant something akin to depression, but held the scope to include frenzy, catatonia, lovesickness and delusions, alongside physical symptoms such as headaches, indigestion and palpitations. In Burton's case, it involved being plagued by feelings of listlessness and despair, choosing solitude and study when he knew that company and merriment would do him good, and taking ten years to complete a degree that he should have finished in four. In 1597, Simon Forman, the notorious astrologer-physician later embroiled in the Overbury plot, treated a 'Robart Burton' for melancholy over five months.[4] We cannot know for sure that Forman's patient and the author of the *Anatomy* are the same person, but the age – twenty years – matches up.

Burton's text gave rise to a wave of interest in diseases of the mind. On a national scale, authors bemoaned the 'distracted' state of a 'world turned upside-down', afflicted by political and religious madness. Playwrights revelled in the dramatic possibilities offered by lunacy; virtually every other character on the Renaissance stage either starts out mad, is driven mad or feigns madness for some cunning purpose. Hamlet manages to juggle all three. But for individuals, there were more pressing questions. Why did some people breeze through life's tribulations while others found every day a struggle? When was religious despair a sign of piety, when a dangerous disorder? Should a person who died by suicide be condemned as a sinner, or pitied for their insanity? Most importantly, how could such maladies be cured?

This chapter is about the people who suffered from mental illness in this period, and those who tried to heal them. Unlike midwives, surgeons or apothecaries, these people did not form a distinct profession. When a person went mad, their friends and family might summon a physician, a vicar or even a musician. Illnesses that manifested in painful emotions or strange behaviour involved the body and the mind, the tangible and the intangible. Their cure required medical skill, human empathy, moral fortitude and, sometimes, a dose of cunning.

* * *

Madness could be a spectacular, frightening, blackly comical affair. As the seventeenth-century physician David Irish observed, mad people could suffer from constant wakefulness and 'a prodigious Herculean strength'. Some endured 'cold, hunger, and stripes [beatings]' without seeming to feel them, or would 'swear, shout, and on a sudden make strange noises...play apish tricks'. The insane were occasionally violent to others, and had no regard for their own safety, often 'pulling their own hair, tearing their clothes, breaking their windows'.[5] They laughed when there was nothing to laugh at, saw things that weren't there and heard mysterious voices. Like Burton's delusional Italian, some mad people held strange beliefs. In 1674, James Newton of Clerkenwell boasted that he'd recently cured a man convinced he was Charles II, who had repeatedly ordered bystanders to have Newton beheaded.[6] Richard Blackmore, who argued that all madness came from disorders of the spleen, encountered a female patient who thought she had frogs living in her guts; his contemporary Nicholas Robinson treated men convinced their bodies had turned into swans, geese or teacups.[7]

These were the most arresting forms of insanity, with which doctors filled their books and later, advertisements. More insidious, however, were the quieter kinds of mental affliction which affected far greater numbers. They had several names, shifting like quicksand as authors sought to describe the indescribable: 'distraction' for people beset by strange beliefs or fears, 'frenzy' for those who turned violent, 'stupidity' for those rendered speechless or catatonic. But the emperor of all these afflictions was the broad and deep category of 'melancholy'.

Melancholy, which kept people of all ages, sexes and ranks in its thrall, was a cruel disease because it was part of the self. After all, the *substance* of melancholy – black bile – was one of the four humours, as essential to the body as blood or air. Melancholic feelings were normal, and Burton reminded his readers that 'no man living is free' from 'dull, heavy, sad, sour' emotions.[8] But to *suffer* from melancholy meant taking the sad and fearful inclinations of that temperament to an extreme, living in constant 'fear of imminent danger, loss, [and] disgrace'.[9] Melancholic people lost their appetite, their powers of concentration and their pleasure in the company of friends. They neglected themselves, for 'the mind, under these horrible ideas…is less attentive to the suffering or affections of the body, being wholly taken up with revolving upon whatever can crowd the mind with the most dark, gloomy, and dismal apprehensions'.[10] In some cases, physicians warned, melancholy could bring about epilepsy and blindness. Without swift intervention, the patient would descend deep into a madness 'where nothing but horror reigns'.[11]

The hinterlands of melancholy were a dark and dangerous place. While few people were recorded as having died of insanity, far greater numbers perished by suicide.[12] In Kent, for example, over three hundred people died at their own hands between 1561 and 1600.[13] Given that the population of that county was only about 85,000, it gives a yearly suicide rate of around 10.5 souls per 100,000 – almost exactly the same as today.[14] These tragedies were not only personally devastating but economically ruinous, because in the sixteenth century, the state cracked down on what it saw as an epidemic of 'self-murder' by making it incredibly difficult to prove that a person who died by suicide had been *non compos mentis* (not of sound mind).[15] In the absence of that saving verdict, the deceased would have their entire worldly goods seized by the Crown, leaving their dependants destitute. As a final insult, they were buried unceremoniously away from church grounds, their graves unhallowed. This is the rule over which two gravediggers squabble in *Hamlet*. Having drowned herself,

Ophelia ought not to be 'buried in Christian burial [ground]', but the coroner is bribed to record her death as accidental.

These draconian rules were designed in part to bolster the Crown's coffers, but they also reflected the moral horror attached to suicide; it was, after all, the one sin that could never be repented, and religious authors continually reminded their readers that a person who impulsively ended their life would spend eternity paying for their decision. The first published text about suicide, John Sym's 1637 *Life's Preservative Against Self-Killing*, described the act as 'the dangerous and damnable practise of diverse persons desperately destroying their own lives, and murdering themselves; with so great prejudice to the honour of God…and with so much hurt to themselves and others', something which was 'in itself evil'.[16,17] But among medical writers, attitudes were more sympathetic. Yes, suicide was immoral, but as Richard Baxter wrote, sufferers of melancholy were so 'weary of their lives' that it was 'as if something within them were either urging them either to drown themselves, or cut their own throats, or hang themselves, or cast themselves headlong, which alas too many have done'.[18] Robert Burton's plea for clemency to self-murderers spoke of a man who had experienced for himself the darkest eddies of despair. When the 'torture and extremity' of melancholy tormented a person, he asked, 'who knows how he may be tempted? It is his case, it may be thine; we ought not to be too rash and rigorous in our censures.'[19]

At the heart of the wild difference of opinion on suicide was the fact that nobody could quite agree on whether madness, and especially melancholy, was a moral, physical or spiritual condition – or even at what point a person should be deemed insane. In 1689, Thomas Tryon, a self-styled health and lifestyle expert, published the long *Treatise of Dreams and Visions*. After the so-called 'Glorious Revolution' which had deposed James II the previous year, the country was once again on the verge of revolt, and Tryon's text capitalised on people's yearning to predict the future. Inside were many claims about dreams and nightmares, including that angels, demons and ghosts communicated with people in their sleep. Tryon also argued passionately that what doctors labelled 'madness' was merely a loss of 'rational faculties' which allowed men to appear in their true, unvarnished state.[20] He pointed out that many of the things 'sane' people did – frittering their estates on prostitutes and wine, fawning on their superiors in hope of promotion, or hoarding money – were far worse than the odd but harmless actions of the insane. 'To speak truth,' he concluded, 'the world is but a great Bedlam, where those that are more mad, lock up those that are less.'[21]

To add to the confusion, it was sometimes difficult to tell whether a person's strange behaviour stemmed from a physical cause, or something more sinister. In the early seventeenth century, physician John Cotta was urgently summoned from his home in Northampton to Warwickshire, to examine the daughter of a gentleman 'of ancient name and seat'. The child exhibited disturbing symptoms. Each day, from mid-afternoon until midnight, she was seized by 'vehement shaking and violent casting forward of her head', accompanied by a bizarre utterance, 'ipha ipha'.[22] Suspecting epilepsy, Cotta reassured the parents that this illness would likely pass, and returned home. Two weeks later, he was called again. Far from remitting, the girl's symptoms had escalated to include whole-body convulsions, intermitted with periods of deathlike stupor. What was even worse was that sometimes the child appeared to be dressing the hair of some unseen woman sat in front of her, as if playing maidservant to an otherworldly mistress.

Cotta did not dispute that supernatural forces existed, though he argued that methods of finding witches – including the infamous ducking stool – were unscientific and unreliable.[23] However, he did not believe this child was possessed. He'd seen many cases of delusions, convulsions and 'fearful and frightful shriekings and outcries', and felt sure this was simply another instance of 'oppressions of the brain'.[24] Nonetheless, he recognised that there was a danger her relatives *would* draw this conclusion, and they could waste precious time on 'cures' such as the witch bottle – a vessel filled with the patient's urine, plus nails, pins, fingernails or clippings of hair, which when buried or hidden around the house would cause the attacking witch's spells to rebound upon them.[*25] He needed to act fast, and in one of the girl's rare moments of lucidity, he persuaded her to utter the name of God. Since possessed people supposedly could not speak in God's praise, this convinced her worried family that she needed medical, not spiritual intervention. With her parents' support secured, Cotta recommended that his patient be sent to Bath to take the healing waters there, and after many months, having lost both speech and use of her legs, she eventually recovered normal functions.

Cotta's *sangfroid* in this case shows that physicians were familiar with mental or neurological illnesses that produced seemingly paranormal effects. In other instances, though, it was the patient

---

* Later, some witches awaiting execution actually confessed to harming the child, which shows that the case must have become locally well-known. Cotta still did not believe that the illness had been supernatural. (*The Major Works of John Cotta*, 170)

themselves who believed that their symptoms were of demonic origin. The most famous of these was a woman whose account of her religious despair, *A narrative of God's gracious dealings with that choice Christian Mrs. Hannah Allen*, is one of the first and most compelling stories of mental illness, told from the inside out.

* * *

Hannah Allen (née Archer) was born in Snelston, Derbyshire, about 1638, and had an ordinary upbringing in an unremarkable Christian household. Aged about twelve, she was sent to live with her aunt and uncle in London to go to school and expand her horizons, and then married the ambitiously named Hannibal Allen in her late teens. It was all entirely normal for a woman of her rank. Except for the fact that, from a young age, she had believed herself to be the most evil person ever to have existed.

Allen was, as she told her relatives, 'persuaded I had sinned the unpardonable sin' and was far from the reach of God's grace, whatever she might be told to the contrary.[26] Quite why she believed this is not clear. The trouble started when after a few years in London, she fell ill and returned to Derbyshire for a period. Whether the illness affected her neurologically, spiritually or emotionally, she became convinced that Satan was putting into her mind 'horrible blasphemous thoughts', and she continued to experience periods of gloom throughout her teenage years and married life, especially when her merchant husband was away for long stretches of time at sea. When Hannibal died in 1664, the dam finally broke. Allen, whose sanity had long hung by a thread, became obsessed with her own spiritual doom. In journals written soon after her husband's death, she confided that 'sometimes the devil tempts me woefully', and salvation seemed impossible: 'I know not what to say…I know not what to do, I shall be undone.'[27] She spiralled into ever more grandiose beliefs about her wretchedness, telling anyone who would listen that 'I was undone for ever; that I was worse than Cain or Judas'.[28] At times she was convinced she would die in the coming hours or days and begged her aunt to bring up her son in the strictest religious discipline, lest he inherit his mother's wickedness. When she was sent to stay at her uncle's house in hopes it might cheer her up, she insisted she would die on the journey, and asked her mother, 'had I not better die in bed?…do you think people will like to have a dead corpse in the coach with them?'[29]

Allen's account of these years is a grim study in the effects of severe melancholy on both the sufferer and their loved ones. On countless occasions, she attempted to end her life, despite being closely watched by her worried family. She sent a maid to buy her opium, thinking she would quietly overdose, appear to have died in her sleep and save her relatives the stigma of a suicide. However, the local apothecaries refused to provide the substance; they likely knew of Allen's condition. Undaunted, she tried other means. She collected large spiders, thought to be poisonous, and smoked them in a pipe, but could not bring herself to consume enough spiders to die.* Once, she snuck out and took a coach

* The belief that spiders – fresh or dried, eaten or smoked – were poisonous was prevalent in the first half of the seventeenth century, though it's unclear why. Spiders were among the poisons supposedly given to Thomas Overbury in the notorious Overbury affair. In 1661, William Ramesey listed spiders in his *Treatise of Poisons*, saying they could cause stupor, swelling of the belly and convulsions. Most bizarrely, James Shirley's 1626 play *The Maid's Revenge* features a villainous apothecary who slips his assistant small amounts of spiders and

eight miles to Barnet Woods, planning to wander until she died of exposure; eventually she changed her mind and returned home. She cut herself with 'sharp sizers' hoping to bleed to death, and in the winter of 1665, as plague ravaged the country, she attempted to starve herself.[30] In her most desperate time, she managed to crawl under some loose floorboards in the family home, lay down and waited to die. It was thought that she had escaped the house and run away until after three days, hunger, cold and cramp forced her to call for help.[31]

With such extreme behaviour, Allen's family might be forgiven for thinking she really was under demonic influence. But instead, they sought medical help, buoyed by a rising tide of scholarship that saw the seemingly possessed as mad rather than bad. In the 1650s, as Allen's 'wretchedness' was beginning, Reginald Scot's 1584 book *The Discovery of Witchcraft* was reprinted. Scot's work argued that self-confessed 'witches' were usually just lonely, uneducated old women whose 'humour melancholical' made them 'confess, that which they never did'.[32] Upon its original release, it was widely decried. Chief witch-hater, James VI of Scotland (later James I of England), ordered that all copies be burned. Yet Scot's view gradually attracted support from physicians and authors, including Robert Burton and John Cotta. On some level, this message had also reached Allen, for even at her worst, she acknowledged it was melancholy humours – sometimes called 'the devil's bath' – which allowed the fiend to slip into her soul.

Patiently and persistently, Allen's relatives sought to show her she was ill rather than wicked. They followed the common

---

other poisons to make him shrink by one inch a day (Wilson, *Poison's Dark Works in Renaissance England*, (Bucknell University Press, 2014) n40–1).

advice for melancholics and provided changes of scenery and company, sending her to London where she took 'much physic' from an apothecary named Mr Cocket.[33] They took her to stay for a week with a minister and his wife, encouraged her to see friends and (in a slightly less enlightened moment) employed a surgeon to let blood from her arm to release the melancholy humours.[34] Nothing worked until 1666, when a family friend, John Shorthose, persuaded Allen's aunt to let her come to him for the summer.

Shorthose was a minister, but his methods were more medical than spiritual, for like his ecclesiastical colleague John Ward, 'he had some skill in physick himself, and also consulted with physicians'.[35] He might have acted merely out of Christian sympathy, or financial motive; residential care was on the rise, and though a man of the cloth was unlikely to run a proper madhouse, there was nothing to stop him taking the odd patient into his home to prop up a meagre income. In either case, Shorthose kept Allen to a strict diet and course of physic, and by the autumn, she had recovered sufficiently to go to church and walk to visit friends. It was the start of an upward spiral; 'as my melancholy came by degrees, so it wore off by degrees, and as my dark melancholy bodily distempers abated, so did my spiritual maladies also'.[36] Allen still believed that Satan had played a part in her disorder, but she now saw him 'working in those dark and black humours, and not from my self'. Provided she could keep humoral imbalance at bay, she could rest assured that she was no worse a sinner than anybody else. In 1668, a decade after the start of her illness, Hannah Allen remarried, to the Warwickshire widower Charles Hatt, 'with whom I live very comfortably'.[37] Her *Narrative* was published in 1683, a frightening glimpse into the depths of madness which nonetheless offered hope for change.

* * *

Allen's account of how her intractable madness was cured is tantalisingly light on details. We don't know what medical regime Mr Shorthose imposed, whether it was harsh or gentle, purgative or nourishing. It seems unlikely that she would have stuck around to endure unpleasant treatment, since on other occasions she fiercely resisted seeing doctors or even talking to friends. One thing is certain: Shorthose's joint expertise in physic and religion was crucial, for medical practitioners of all kinds broadly agreed that melancholy illness unbalanced both body and soul. One of the most lucid explanations came from Lazare Rivière, a famous French physician who attended Louis XIII.[*38] He explained how the connection between brain and body relied on 'animal spirits': a mysterious substance generated in the blood, refined enough to travel from the brain through the nerves to enable sense and motion. When a body had too much melancholy humour, those spirits – usually 'pure, thin and transparent' – became cold, dark and sluggish, producing 'sorrow and fear' which clouded rational thought.[39] Thomas Nashe (the author who'd imagined a greedy anatomist cutting up hapless travellers) put it in grosser terms. Like stinking sulphuric gases emanating from the bowels of the earth, he wrote, so melancholy humours rose from the spleen

* Rivière's text was translated by the workaholic apothecary Nicholas Culpeper, and printed by his friend Peter Cole.

to 'intoxicate' the brain. There they bred 'fearful visions' in the imagination, trampling the helpless intellect underfoot.[40]

Melancholy struck at the heart of one's humanity, and its causes were frustratingly mysterious. The connection between thoughts and humours could work either way around; for example, being frequently angry would increase the body's production of the hot humour, choler, but an excess of choler would also prime a person to become enraged at the slightest provocation. This meant that potential triggers for melancholy illness lurked around every corner, and in Burton's compendious work on the subject, over a hundred pages were taken up with various causes. There were the usual suspects of bad diet, exercise (too much or too little), bodily evacuations (too often or too seldom), unwholesome air, ill astrological influences, old age, inheritance from one's parents or ingestion through a nurse's milk.[41] But added to these was the raft of misfortunes which formed part of every life: insults, illnesses, frightening sights, poverty, bereavements, marital discord, inability to conceive or simply a trial sent from God. Burton, like many others, singled out harsh educational practices and excessive study as particularly likely to wreak mischief.[42] Corporal punishment was routine in the Renaissance schoolroom, and to succeed as a scholar meant long, lonely hours memorising classical tragedies and dead languages. The image of the melancholic poet was a cliché, but it existed for a reason.

Like Allen's family, most authors on melancholy took a sympathetic approach, suggesting common-sense lifestyle changes far removed from the punitive treatments of later centuries.*

* In the eighteenth and nineteenth centuries, it became increasingly common for physicians to 'treat' mentally ill patients with beatings and immersion in cold water.

As always, the first step was diet; plenty of hot and moist meats like rabbit and wildfowl to counteract the chilling, drying effects of black bile. Hot baths, usually eyed with caution, were considered good for melancholics' cold and dry complexion. Various plants such as violets or camomile might increase the effect (though Burton also suggested adding a boiled ram's head to the tub, which was potentially less relaxing).[43] Moderate exercise was good, especially hawking, fishing (Burton's second-favourite topic) and riding, which all had the added benefit of getting the patient outdoors. These measures were thought to have a physical effect, counteracting the troublesome melancholy humour. But other recommendations show that physicians approached melancholic patients in a holistic way, seeking to lift their mood and divert their thoughts through what we might now see as occupational therapies. Lively music was often recommended; John Ward wrote that 'many learned and credible authors' attested how 'music cureth the brain of madness and the heart of melancholy'.[44] The afflicted person should have 'cheerful sights' to look at and be kept out of the dark, which 'is as it were a pattern of death'.[45]

Changing a person's environment was simple, but how could physicians redress disordered thoughts? Burton advised that friends of a melancholy person should undertake a talking cure, asking them: 'thou art discontent, thou art sad and heavy, but why, upon what ground?…Examine it thoroughly, thou shalt find none at all'.[46] (Evidently this rousing speech was not foolproof, for Burton added that friends could also try jabbing the patient with something sharp, to jolt them out of their malaise.) For love melancholy, he had a raft of far more colourful suggestions to quell the sufferer's ardour:

> [Tell them] she is a fool...a slut [slattern], a vixen, a scold, a devil...that he or she hath some loathsome filthy disease, gout, strangury [difficulty urinating], falling sickness, the pox...that she is bald, her breath stinks, she is mad by inheritance, and so are all the kindred...That he is an hermaphrodite, a eunuch, imperfect, a spendthrift, a gamester...a common drunkard, his mother was a witch, his father hanged, that he hath a wolf [cancer] in his bosom, a sore leg, some incurable disease, that he will surely beat her, that he walks in the night, will stab his bedfellow, tell all his secrets in his sleep, and that no body dare lie with him, his house is haunted with spirits.[47]

Burton's list may say more about its author's hang-ups than the likely hazards of love, but the general principle held true. These were psychological measures before the science of psychology existed, based on an understanding of the human condition rather than human anatomy. If they failed to work, however, there were more extreme routes available. Shaving the unwell person's head was common, not to humiliate them but to cool the fevered brain. In 1702, the *Post Man* newspaper ran a poignant missing persons notice for 'a little fair woman' who had run away from home while suffering from melancholy. She wore a 'sad coloured stuff gown and petticoat' and a 'sky coloured' under-petticoat, but was made distinctive by her shaved head, which was bound with white cloths.[48] Bloodletting, which Hannah Allen underwent, was supposed to relieve madness of all kinds by evacuating the troublesome humours. While most practitioners simply opened a vein and let the blood into a bowl, Burton noted that leeches were particularly good in relieving melancholy, especially if they were applied to a person's haemorrhoids, where they sucked the foul humours away from the brain and heart. If this was not sufficiently unpleasant, one could achieve a similar effect by burning the patient with a hot iron.[49] Purges of various sorts supposedly worked on the same basis, though in reality they probably just made the mad person too weak to carry on with any outlandish behaviour. Altogether, it was a therapeutic free-for-all. The protean symptoms of melancholy meant that virtually any substance could be tossed into a medicine to cleanse or strengthen the body, from the innocent (herbs, flowers, tree barks) to the hazardous (hellebore, laudanum) or downright bizarre (a wolf's heart, to be carried around or eaten).[50]

For the most part, medicine worked in much the same way for madness as for any other condition. People called in a practitioner they trusted and could afford, took their advice and then cared for their sick relatives at home. Ever alert to new opportunities, certain seventeenth-century physicians and surgeons started to specialise in the cure of the insane, using the new medium of newspapers to advertise their services. Some offered mail-order cures, like a 'digestive powder' which promised to cure even those patients who were 'crazed and almost distracted' for the low price of three shillings and sixpence.[51] The most enterprising, however, recognised that diseases of the mind could be more frightening and puzzling than those of the body. They saw what families of mentally ill people really craved was peace of mind, someone to shoulder the responsibility of keeping their troubled relatives supervised and safe. Once again, the fertile soil of the marketplace yielded a new crop of medical practitioners – the asylum keepers.

* * *

The rise of private madhouses was almost unnoticeable at first. From the sixteenth century, physicians sometimes took physically ill people into their own homes to offer more intensive care, and likely some did the same for the odd client who was mentally unwell. The state had long provided hospitals for the impoverished insane, including the infamous Bethlem Hospital, better known as Bedlam. But our first glimpse into a private madhouse dates from about 1606, when a young gentlewoman called Dionys Fitzherbert was admitted to the care of 'Doctor Carter', in the Holborn area of London. Like Hannah Allen forty years later, Fitzherbert was deeply afflicted by thoughts of her own wickedness, and the conviction that she would soon be burning in hell. In letters and diaries, she wrote of being a 'loose liver and a thief'. She was suicidal, suffering 'fearful tremblings and astonishments', sure that at any moment, her incomparable sins would be exposed to the world.[52] Soon, she veered from religious enthusiasm to full-scale delusions: that her relatives did not know her, and would have her killed for pretending to be related to them; that she would vomit enough to cover the whole of Charterhouse Yard; that her doctor was God; that there was no such thing as death.[53] Her family tried shutting her in her room for four weeks; to their surprise, this made her worse, and it was then that she was moved to Dr Carter's establishment. That she was given physic, was attended to by a female 'keeper' and shared the space with other 'company' suggests that this was not simply the physician's private residence, but an early – maybe the very first – private madhouse.[*54]

* It proved an effective remedy. Fitzherbert returned home after a few weeks and later composed an account of her 'afflicted conscience' which she deposited in Oxford's Bodleian Library.

Such places remained rare in the early seventeenth century. Helkiah Crooke, whom we will meet again later, was one of a few who provided private care for well-heeled lunatics. In 1630, he arrived in a coach, flanked by three strong men, to apprehend the Bedfordshire gentleman Edmund Francklin and forcibly take him back to London. Francklin had been 'found a lunatique', and his family paid Crooke a handsome £200 a year to provide him with clothes, lodgings, food, constant supervision and unspecified 'physick' until he was cured (there is no record of whether the cure succeeded, or how long it took).[55] Gradually, private hospitals for the mad became less of a rarity, especially in London. In 1643, a pro-Parliament text entitled 'A Looking-Glass for Malignants' told stories of the various terrible fates that had befallen Royalist 'God-haters', from accidents and murders to the birth of a child with no head, but eyes, ears and a mouth all located in its abdomen. One of the less fantastical tales concerned a hosier (maker of stockings) who became 'distracted and besotted in his senses'. The local authorities 'held fit to have him away to Bedlam' before his friends intervened to ensure 'he was not put into the common condition of the madmen there, but was kept private in the house of one that endeavours the cure of such persons'.[56]

Clearly, by this point the horrors of Bedlam (that is, Bethlem Hospital) were well-known, and private care had emerged as an alternative to that most feared of institutions. But it was only after the Civil War of 1642–51 that madhouses really began to multiply. When the bloody tide of conflict receded from the country, it left behind a flotsam of shattered minds and bodies. Some of the physically broken were mopped up by a growing network of hospitals – though plenty also ended up on the streets, surviving as best they could. For the mentally ill, however, private asylums took in both those who could afford to pay, and those paid for by their local parish. In the advertisements for such places, they appear as something between a hospital and a spa hotel, where kindly staff feed patients nourishing meals whilst they recover under the close attention of a skilled physician. David Irish owned two madhouses in Surrey – one at Stoke near Guildford and another near Egham which was managed by his son. He promised prospective patients a daily meat dinner, 'good fires', and 'decent chambers', 'far beyond what is allowed at Bedlam'.[57] Others prided themselves on taking supposed lost causes. James Newton claimed that in his madhouse at Clerkenwell, he'd cured 'some that have been twelve years mad and others that have been sixteen years melancholy', and in record time too; 'the last ten I had went away in a short time very well, except one that was two, another three, and the third four months…several others have been cured in a week, or very few weeks' time'. As a final flourish, Newton offered to take three patients out of Bedlam and cure them gratis – Bedlam did not take him up on the proposal.[58]

These were comforting thoughts for families of the insane. Sometimes, they may even have been true. However, this was the Wild West of patient care, a new, completely unregulated industry in which anybody could set themselves up as a mad-doctor without heed to such trifling matters as medical expertise. The consequences could be grave. In 1661, Reverend John Ashbourne was running a small medical practice in the village of Norton, Suffolk, specialising in the treatment of the deranged. He needed to top up his income, for the Civil War had badly disrupted the tithe system on which he and other ministers relied for their upkeep, so he decided to start using the rectory – his family home – as a small-scale asylum. There were six bedrooms upstairs, plus a living room downstairs that Ashbourne converted into a bedroom, and between them they held thirteen beds – enough for Ashbourne and his wife Abigail, their children and their servants with room left over.[59] For fourteen shillings a week, plus two lump sums of £10 at the beginning and end of treatment, Ashbourne would take in 'distracted' people along with one of their servants and supply them with food, laundry and everything else they required until they were well enough to return to their normal lives. He described himself as a 'practitioner in physic', but it is unclear what, if any, treatment he supplied to the people in his care, and his inexperience would soon prove fatal.[60]

On 1 August, Ashbourne was on his way to visit a patient, making the most of the fine summer weather as he took an easy walk through the flat countryside. It was hay-making season, and in the meadow lay rows of cut grass to be turned and dried, with the labourers' tools marking their progress. Ashbourne was met on the way by one of his patients, Mr Ward. The exact identity of this client is uncertain, but it was probably Samuel Ward, Ashbourne's own brother-in-law. The records show that *something* was certainly different about Samuel, for despite being an eldest son, he did not inherit on his father's death, but was allotted £20 a year with which his brothers Nathanial and Joseph would make sure he was kept 'in a comely and decent manner for and during his natural life'.[61]

Ward was a resident in the madhouse but had been doing so well that he was allowed to come and go freely, without a keeper. Or, at least, he seemed to be doing well. Without warning, Ward grabbed a pitchfork from the hay pile and ran it straight through

Ashbourne's neck. He then took the wounded man's own knife from his pocket and stabbed him seventeen times.[62]

Ashbourne died at the scene, but Ward's fate remains shrouded in mystery. He was carted off to Bury jail, but there is no record of his trial, so he was likely deemed *non compos mentis*. Maybe he was put into another private madhouse, or into Bedlam – though the latter was supposed to house only curable patients, the fact that the hospital later built a wing for the 'criminally insane' shows that in practice it sometimes became a dumping ground for those who could not be kept anywhere else. Surprisingly, however, the reverend's murder did not spell the end of the Norton madhouse. Ashbourne's widow Abigail had long been integral to the family business, but she needed the legitimacy of a male practitioner at the helm to continue practising. Luckily, the couple's son Samuel already had a BA from Cambridge, and only ten weeks after his father's death, he successfully applied to Trinity College for a medical licence. Whether Samuel actually did much at the madhouse is a different matter. He died in 1685, yet in 1686, a local gentleman left £100 in his will to pay for the upkeep of their 'distempered' relative in the Ashbourne house. Abigail – by now an old woman, first widowed, then bereaved of her son – was running the asylum herself.[63]

The madhouse-keeper: John Westover

Most of the seventeenth century's madhouses have faded into obscurity, the records of their patients long lost. We see them only in momentary flashes, as when Ashbourne's murder drew the attention of the press. But in a ragged volume from a sleepy corner of Somerset, we can view a private asylum as it evolved over years. This is the casebook of John Westover, surgeon and mad-doctor. A long, narrow book, it is the only survivor from the manuscripts Westover kept throughout his life, and along the way it has lost its covers and some of its paper. In the 452 pages that remain, Westover recorded his day-to-day business from January 1686 to October 1700.

Born in 1643, amid the turmoil of war, John Westover was named after his father (1616–79), a first-generation surgeon from a line of farmers and yeomen. The Westovers lived in Wedmore, a small village which occupied a high point in the sodden landscape of the Somerset Levels. Each winter, swathes of the surrounding county were submerged despite the network of rhynes (ditches) that criss-crossed the landscape; in 1607, relentless rain and the high spring tides of the Bristol Channel caused a flood that swallowed up 200 square miles of farmland and killed over 2,000 people. Like many of his colleagues, Westover junior was not too fussy about the official boundaries of his business. The scrutiny of the College of Physicians did not reach rural Somerset, and besides many abscesses, sores, broken limbs and farming injuries, he did brisk trade treating worms, fevers, malaria and scabies, a parasitic skin disease which affected a quarter of his patients. The first entry in the book records how Westover attended William Counsell from the neighbouring village of Sand when he fell and sustained a compound fracture in which 'the [thigh] bone came through the skin'.[64] Four days later he visited Richard Salter's wife a few miles away and gave her a rhubarb purge to expel the worms that had been making her sick for a fortnight – he noted in the book that Salter had not yet paid his debt.[65]

Alongside this everyday practice, however, Westover also gained a reputation for curing illnesses of the mind. His casebook lists fourteen men and forty-four women whom he treated as outpatients for 'melancholy', 'hysteria' or 'distraction', using the traditional methods of purges and rebalancing medicines. On 13 June 1686, Mary Counsell paid Westover half a crown for a purge for her 'mellencholey'. The next year, he gave Thomas Coomer a 'long-necked bottle of julep [sweetened medicine]' for his melancholy sister, advising her to take two or three spoonfuls twice daily.[66] Grace Poole suffered from hysterical fits so violent that it took three people to restrain her. He sent her medicines worth three shillings, including 'oil of amber' to put around her nostrils (an early equivalent to smelling salts).[67] Westover's methods contained nothing new, but they gave comfort to his clients, and many returned year after year. Whole families trusted themselves to his care, and in return he gave easy terms, allowing Thomas Millard's wife to delay payment until she'd sold her cow and occasionally writing off debts when he knew the family couldn't meet them.

As Westover's reputation grew, people began to seek him out from further afield, and business flourished despite competition close to home; Glastonbury, nine miles from Wedmore, possessed an asylum which had been in operation since at least 1656. (This was periodically home to Reverend George Trosse, admitted when he heard and saw devils – often disguised as items of food and drink – telling him to kill himself.)[68] In 1680, the year after his father's death, Westover decided to take the plunge and expand his offering by opening his own madhouse. After all, he could offer space, some medical expertise and as much fresh air as the patient could handle. He occupied the handsomely porticoed

'Porch House', and, in 1680, he added a home office of massive proportions – a three-storey building with a staircase running up the outer wall, just twenty yards from his own residence.* Here, patients who needed longer-term or more intensive care could be accommodated. Not all the residents were mad poor Ann Harvey was in for eleven days for the triple misfortune of dropsy, scurvy and a tooth extraction – but the building became known locally as the madhouse for a reason, and through the clues in Westover's accounts, we can peek at the lives of those confined within.

As soon as a person passed through the thick stone doorway into the 'hospice', their bill began to mount. Most of what Westover charged was for 'tabling', or bed and board. It cost five or six shillings a week to 'table' a resident, the equivalent of two days' wages for a skilled tradesman like a blacksmith. Though the sum wasn't particularly steep, many of Westover's patients were with him for months or even years, and the expenditure added up. In 1699, for instance, Alice Stevens was admitted for 'madness', and by the time she left 'parfitly cured and well of her distemper' thirty-one weeks later, she'd accrued tabling costs of £5.** Then there were the costs of actually treating the patient,

* This also happens to be the village where I lived as a teenager. Unbeknownst to me, my parents nearly purchased Westover's old house as a family home in the 1990s, but my mother was concerned about being haunted by ex-patients – a reminder of just how long the stigma of the madhouse sticks around.

** Being 'poor', Alice had no prospect of paying for herself; her stay was funded by the local parish, and Westover's distrust in them is evident from the fact that he recruited three witnesses to sign a memorandum attesting that Westover had only received part payment on Alice's departure.

mostly with unspecified 'potions'. Many clients chose to pay a flat fee for the cure of their relatives. William Williams sent his wife Mary to Westover in December 1698, having agreed that he'd pay £15 for her cure if it was done by Christmas, and an extra £5 if she needed to stay longer. It was an unusually large sum, which suggests that either Williams was very wealthy or Mary was so unwell that she needed extra care. In the event, Mary's son came when Westover was out and removed his mother after just eleven days, leaving only a third of the agreed fee.[69] Negotiations between Westover and the families of his patients were often difficult, and never more so than when George Vowles, who'd been admitted for 'distraction', died after six weeks in the hospice. What caused George's sudden death is a mystery – he'd been confined in the hospital twice before without incident – but Westover was clearly worried about the impact on his reputation, because he allowed George's father Robert to pay a fixed fee for his tabling and 'please his self' when it came to paying for the medicines (only fair, since they clearly hadn't worked).[70]

To weave these individual threads into a tapestry of patients' experiences, we must read between the lines of Westover's carefully kept accounts. This establishment was a small one, housing only a few patients at a time, and given that it often allowed payment in jewellery, livestock or cheese, it evidently wasn't as slick a business as some of those advertised in contemporary newspapers, where the great and good sent their embarrassing and inconvenient. What was more, the hospice was no sealed unit, and much of a patient's experience depended on the attentions – or inattention – of their friends and family. When Mrs Silver, the wife of Captain Thomas Silver, entered the madhouse in 1700, her stay was positively merry at first. For reasons unknown (though maybe guessable), male callers – Thomas Hill, John Biggs, William Bond and her husband's younger brother – visited her frequently, bringing brandy, tobacco, raisins, anchovies, capers and copious quantities of claret. But in 1701, Mrs Silver's husband called John Westover to a meeting at his home near Bridgwater, and the visits abruptly ceased. In fact, she received no further visitors or gifts until she was discharged in 1705. Captain Silver might have been punishing his wife, or he might have been keeping her safe from predatory men, but the result was that she spent four long years away from home with no one but the doctors and her fellow inmates for company.

Westover's longest-staying resident was John Edwards, the youngest son of a gentleman from nearby Mudgley. In 1660, Edwards was listed in his father's will (witnessed by Westover senior) with no indication he was any different from his brothers. Seven years later, however, something had gone badly awry, for his mother noted that she had paid £5 to a 'Mr Wester' – probably Westover – for 'keeping and governing' her son for sixteen weeks.[71] For nearly twenty years after this incident, Edwards was calm enough to live at home, but his brother Richard died in 1687, and as his family fell apart, so did Edwards's sanity. He was admitted to the hospice in 1687 for around nine months, and then discharged, but by 1689 he was back in confinement, and in 1690, his mother died – a loss from which he would never recover. The family estate was taken over by a distant relative, who did not know or care about his lunatic kinsman. John's brother had made provision for him in his will, but nobody was left to look out for him in more than a financial sense. He spent the next seventeen years until his death in the madhouse, forgotten by the outside world, and the only hint at his treatment there comes from one of the few things Westover purchased for his care – a length of rope.

* * *

John Edwards's experience shows how the market in madness changed the lives of mentally ill people for better and worse. The new interest in melancholy, spurred on by medical texts and personal accounts, shone a light on mental disorder. Even in the depths of rural England, there was now little danger of Edwards being viewed as possessed or under the influence of witches. Though fears of the occult lasted for many decades to come, the last people executed for witchcraft in England were Temperance Lloyd, Mary Trembles and Susannah Edwards in 1682. They closed a chapter in which around five hundred men, women and sometimes children had been tortured and killed for crimes they did not commit. The medicalisation of mental disorder also helped snuff out the legal penalties – if not the social stigma – that attended suicide. By the 1700s, over ninety percent of suicides were judged to have been *non compos mentis*, and the brutal practice of seizing goods from a deceased person's family was almost obsolete.[72]

But where the market shone light, it also cast dark shadows, for there was now money to be made from the insane. Despite the rosy picture painted by asylum keepers, stories from their patients reveal that in the tussle between medical best practice, the wants and needs of paying relatives, and the madhouse's business model, profit usually came out on top. Edward Trevor, who was sent to a madhouse by his mother in the 1670s, recorded how one of his front teeth had been broken during an attempt to force-feed him, and he was later chained to the floor.[73] James Carkesse, who was admitted to Thomas Allen's madhouse in Finsbury in 1678, alleged that despite his enlightened reputation, the doctor was a 'mad-quack' who knew no medicine but purging, bleeding and vomiting, and chained Carkesse up in the dark.[74] Worse still, scandalous tales sometimes emerged of sane people incarcerated by villainous relations. In 1694, for instance, James Newton – the man who'd offered to take on three Bedlam residents – had to release his patient Lady Honoria Willoughby after she and her friends alleged that she had been set up by her 'ungrateful, false, and covetous' husband.[75] Willoughby was well-connected and articulate. Others may not have been so lucky.

In 1774, a law was finally passed decreeing that all madhouses had to be licensed by the Royal College of Physicians and inspected at least once a year by the local courts, with patients admitted only by medical certificate (the first 'certifiably' insane). By then, it was like applying a sticking plaster to an open wound, for institutions of varying size, proficiency and humanity were proliferating all over England. The era of the asylum had begun, and it would not end for nearly two hundred years.

A troupe of travelling performers including a toothdrawer. Oil painting, seventeenth century, after Theodor Rombouts. The Wellcome Collection.

CHAPTER SEVEN

Those They Called Quacks: Unauthorised Healers

It is a cold and windy market day in Northampton, and the quacks are out to ply their trade. They start early, hanging banners and flags with roughly painted pictures of mangled limbs and faces to make their business known to those who cannot read. Opening their bundles and packs, they lay out rows of bottles and parcels, wooden false legs, harnesses to truss hernias and straighten hunched backs. When a customer draws near, the patter begins, 'discoursing with what agility they can solder new gristles for old noses, and newly again infranchise French [paralysed] limbs, and finally making themselves admirable tinkers of all infirmities'. Some of the most determined will even appear to cut their own flesh, discreetly piercing a chunk of liver or a packet of pig's blood secreted up their sleeve. When the gore is wiped away hours later, onlookers who saw a wounded man at the start of their morning's market business will witness the 'healed' flesh. Maybe they will be convinced to buy a little of the potion that effects this wondrous cure.

Watching these spectacles from a distance is John Cotta, the physician who treated the Warwickshire gentleman's catatonic child. He notes the pinched and hungry faces on some of the traders, men who claim to be selling medicine laced with gold (*aurum potabile*) yet cannot muster up a penny for a loaf of bread.

He is unsympathetic. Cotta loathes them and scorns their clientele. Soon, he knows, the quacks will pack up their wares and move on to the next town, and he (thinking little about his own steep fees) professes himself baffled at how in every place, these creatures of 'idleness and theft and beggary' manage to convince otherwise sensible men and women that they are touched with a spirit of revelation that has rendered them gifted physicians without the need for formal training. In the end, he sneers, 'like noble chemists, having extracted silver out of the baser metal of idle words, in smoke they vanish, leaving behind them the shadow of death'.[1]

* * *

John Cotta was a sceptical man, little given to credit tales of witchcraft and wonders. Most of his friends and patrons in the parish of All Saints, Northampton, were staunch Puritans, buttoned-up folk who stood stiffly against all forms of disorder and disrepute. So it vexed him when the rules of the medical profession were broken by those he considered 'ignorants', 'quacksalvers, bankrupt apothecaries, and fugitive surgeons', travelling the country to 'creep into the favour of mean people', selling wares which carried the titles of famous medicines, but on the inside were 'infamous poisons'.[2] Unluckily for Cotta, he lived in a time when such 'itinerant' practitioners – that is, those outside the medical establishment – were flourishing as never before.

In two books, *A Short Discovery of the Unobserved Dangers of Several Sorts of Ignorant and Unconsiderate Practicers of Physick in England* (1612) and the more snappily titled *A True Discovery of the Empiric with the Fugitive, Physician and Quacksalver* (1617), Cotta introduced a rogues' gallery of unlicensed medical practitioners whose activities spanned the length and breadth of the country. There were 'empirics', men who learned their trade by experience and experiment (empiricism) rather than study; 'heretic physicians', who strayed dangerously from Galenic methods and tried their own path; 'practicers by spells', who claimed to cure diseases by magic; and witches and wizards, some of whom actually possessed the power to heal and harm. 'Fugitives, workers of juggling wonders, and quacksalvers' mixed cure-alls with showmanship, selling their dubious wares by dazzling their audiences. 'Servants of physicians' picked up snippets of knowledge from their masters but were soon out of their depth when they tried to cure more than the simplest diseases, while the whole category of women healers was found by Cotta to attain at most an 'allowable mediocrity', and then only rarely.[3] Vicars and parsons did not escape Cotta's scrutiny. Though himself deeply religious, he resented the fact that clergymen (namely, his local parson John Markes) took business away from proper physicians (namely, himself) by selling their own remedies, and he devoted pages to denouncing their 'rash, ignorant, and unskilful errors and commissions against the health and life of many'.[4] Even legitimate branches of medical study quickly turned from fair to foul; astrologers overstepped their bounds by taking up prognostication, and 'conjectors by urine' thought their method infallible when it was a small aspect of the true physician's craft.[5] Indeed, Cotta complained, there was nothing under the heavens that could not be made an instrument of 'adulterate seeming, lying, and cozenage'.[6]

Cotta and his colleagues bundled all these many undesirables into one basket labelled 'quacks' – the word 'quacksalver' derived from the Dutch *kwakzalver*, a hawker of salves and ointments. Such people were also commonly called empirics or 'mountebanks', from the Italian *montambanco*, meaning a person who climbed on a bench to shout about their wares. It was not only other medical practitioners who obsessed over these characters. In the fevered imagination of the era's writers, they were filthy foreigners, hailing from southern Europe and bringing with them the follies of their native lands. In 1605, Shakespeare's friend and rival Ben Jonson produced *Volpone*, a raucous tale of an Italian trickster who deceives his greedy hangers-on into believing he is about to die and leave one of them a huge fortune. Among his many japes, Volpone disguises himself as a mountebank and bamboozles the crowd into buying vials of oil 'guaranteed' to cure an increasingly outlandish list of diseases:

> cramps, convulsions, paralysis, epilepsies, tremor-cordia [palpitations], retired nerves, ill vapours of the spleen, stopping of the liver, the stone, the strangury [stoppage of urine], hernia…iliac passion [appendicitis]; stops a dysentery immediately; easeth the torsion of the small guts: and cures melancholia hypocondriaca [hypochondriac melancholy], being taken and applied according to my printed receipt.[7]

It is, of course, all pomp and puffery, assisted by the troupe of hermaphrodite dwarves Volpone keeps for his own entertainment.

This is the popular image of the quack: a villainous, flamboyant, travelling shyster. Jonson's depiction played on every fear the English had about their glamorous Popish neighbours, and on the wider suspicion of people who travelled about, potentially spreading both disease and dissent. Yet even Cotta admitted that unlicensed practitioners varied vastly in their methods and skills. Some moved from town to town, some owned their own premises, others served only their local community, still others sold goods through agents, or by post. Even among the much-distrusted itinerants, there were some whose long experience made them supremely adept, capable of rivalling any physician in their narrow speciality.

A prime example of skilled quackery was 'couching', an operation to remove cataracts. It involved piercing the limbus, the point where the cornea meets the white of the eye, with a sharp needle or thorn. The cloudy cataracted lens was then dislodged from underneath, so it remained in the eye but was no longer obscuring light from entering the pupil. The patient would not regain their vision, since the lens could not be replaced, but if the operation was successful, they would be able to discern light and shade – provided they did not get an infection which might mean the loss of the eye altogether. Couching was not technically difficult; it was mentioned somewhere in most books of surgery. But it required experience, strong nerves and a steady hand. A deft itinerant with hundreds of operations under their belt might well be more skilled – not to mention far cheaper – than a scholarly ingenue fresh out of medical school.

It was a similar story in many specialities. An awkward fracture could be realigned in seconds by an experienced bone-setter, who might be found in most provincial towns. With the help of another strong man, they would pull the two parts of the broken limb apart before replacing them in their proper alignment (much the same as today's practice, though without the benefit of X-rays or anaesthetic). They then applied bandages stiffened with wax, followed by splints wrapped in cloths wet with rosewater.[8] Often the bone-setter and tooth-drawer were one and the same, for they both required a combination of skill and brute strength. Many of the same people who couched for cataracts also operated on cleft lips, so that in 1715, 'Her Majesty's Oculist' William Read was obliged to take out an advertisement warning that men roaming the country claiming to be his servants were actually villains exploiting his good name.[9] Even lithotomies (bladder and kidney stone removals) were carried out by travelling practitioners, who drummed up trade by claiming to get the ordeal over with faster than anybody else.

Itinerant surgeons were the most controversial of all irregular practitioners, much maligned by 'authorised' practitioners. However, what all these 'shameless' people knew – and Cotta refused to accept – was that quacks flourished because they filled a need, supplying a growing populace with cheap, quick, sometimes effective treatments. Untrammelled by tradition (or regulation), an industry grew rapidly, reflecting and distorting official medicine like a hall of mirrors. Some characters, like cataract surgeons, almost matched their licensed counterparts; others, like mountebanks, twisted the figure of the physician into something stranger. Soon, we will meet faith healers, cosmeticians and prosthesis makers, whose resemblance to medical professionals was fleeting and fragmentary. But before their time was to come, the market would be upended once again.

***

It is easy to imagine medicine – and medical progress – as an edifice built entirely on the triumphs and failures of individual men and women. One genius could carve out an entirely new field of scientific investigation; one miscreant could bring a system to its knees. We've met some of both these archetypes. But medicine was also the product of an ecosystem in which apparently distant global events could produce ripples which spread to affect the health of ordinary English patients. This was the case of medicine-by-post.

In 1618, a group of Protestant nobles threw three Catholics – Count Vilém Slavata of Chlum, Count Jaroslav Bořita of Martinice and their secretary Filip Fabricius – out of the window of Hradčany Castle in Prague.* This event, which became known as 'the defenestration of Prague', had many causes, for continental Europe was a roiling cauldron of political and religious power struggles. Chief among them, however, was the Kingdom of Bohemia's new ruler, Ferdinand of Styria, a man distinguished by his magnificent curled moustache and devout Catholicism. Where previous rulers of Bohemia (the modern-day Czech Republic) had been tolerant of the region's religious mix, Ferdinand was anything but, and set about eroding the religious freedoms of Protestants and muscling in on the wealth of Protestant nobles. The violence that resulted set alight a tinder box of religious tensions on the continent, and started a war which would last thirty years and claim nearly eight million lives.

* Despite falling seventy feet, the trio did not die. Catholics claimed this was due to divine intervention, while Protestants claimed they had made a soft landing in a dung heap.

This mattered for medicine. It was important because, as we have seen, innovations in surgery accelerated when there were so very many wounded to practice on. It also sent a flood of Protestant refugees to England, including medical practitioners eager to share their skills, like the Dutchman Joannes Groenevelt.[10] But the war also had a less obvious effect on medicine: it gave a kick-start to the development of the newspaper industry.

With conflict spreading on the continent, the English appetite for printed news skyrocketed. By 1618, more people could read than ever before, printing was getting cheaper and censorship was far more relaxed than under the paranoid years of Elizabeth's reign. News entered the country first in a trickle, then in a flood, and enterprising printers such as Nathaniel Butter and Nicholas Bourne were quick to jump on the bandwagon, starting their *Certain News of the Present Week* in 1622. Soon enough, it was clear that newspapers were the new word-of-mouth, supplying a growing populace with hard political news, and also gossip and tales of wonders. The *News Published for Satisfaction and Information of the People*, for instance, had reports from Paris, Moscow, Rome, Amsterdam and the slightly less glamorous Yarmouth (trouble on the Barbary coast, a huge fire, the Pope's gout, plague and fears of war respectively) alongside advertisements for new books, a powder to treat epilepsy and false teeth.[11]

In the turmoil of the English Civil War, partisan newspapers drew the battle lines for each side, claiming victories and each accusing the other of atrocities. Skipping neatly between one side and the other like 'a politick shuttle-cock' was a man named Marchamont Nedham.[12] Nedham was a qualified physician, but when war broke out he saw the chance to make his fortune as a journalist and newspaper editor. Happily unburdened by principle, he switched allegiances according to whichever way the political wind was blowing (though not quite swiftly enough to avoid landing himself in jail twice over). His contemporaries painted him as 'a man of low stature full set, black hair, hollow-hearted, empty skulled'.[13] But what Nedham lacked in conviction, he made up for in shrewdness, and he saw that newspapers could make far more money if they sold advertising space rather than relying only on shifting copies. As a once-and-future physician, and a vocal anti-Galenist, it was only natural that he turned to medicine-sellers to fill his empty pages.

By the 1650s, newspapers could be found in every coffee-house, passed between the patrons and read aloud over cups of steaming coffee or chocolate (women took their chocolate at home; it was thought dangerously aphrodisiacal).* Between military reports, high-society gossip and horse-racing results were stuffed dozens of advertisements, many for medical goods of all kinds. Nostrums claiming to cure everything from gout to jaundice were described in rapturous terms. A 'famous water' known as 'Aqua-anti Torminalis' (that is, 'anti-death water') was said to be 'incomparable' against colic and wind, 'having in a little time restored several to the use of their limbs, which had been taken from 'em by such terrible distempers'.[14] The same paper contained advertisements for: a physician selling pills to cure the clap; a diet-drink for kidney pains and a mouthwash to whiten the teeth; a 'never-failing ointment' to get rid of body lice; 'Stoughton's elixir stomachicum' to cure nausea; a 'tincture

* Chocolate was also thought to be an effective hangover cure, and was drunk by Pepys the morning after Charles II's coronation in 1661.

of metals' to remedy apoplexy, epilepsy, melancholy and fits; and the latest edition of Hannah Woolley's *The Accomplish'd Lady's Delight*.

When newspapers first began, novelties like these were only accessible to Londoners, or at least those in large towns. In the absence of street maps, advertisers directed potential customers by landmarks – near St Paul's, next door to the butcher, round the corner from the pub. But this would soon change. In 1635, Charles I had given his top postmaster Thomas Witherings the authority to establish a postal service stretching to all England's major cities, and in 1680, the penny-post came into being. Letters and small parcels could now be sent from hundreds of 'receiving houses' in central London (many of which were, conveniently, also coffee-houses) to other London destinations for next to nothing. Sending to locales near London, such as Hackney, was just two pence, and the larger post system now spanned the entire British Isles.

This new model was a revelation for both customers and medicine-sellers. Patients in the provinces were no longer bound to the advice of their local medical practitioners, for if they could access a newspaper, they could access dozens of medicines at incredibly low prices. The cures advertised in the *London Post* all retailed for just a shilling. Medicines-by-post also marked a fundamental shift in how people thought about healthcare. For hundreds of years, the Galenic model had dictated that all illnesses came from disruption of the humours, which meant that medicines needed to be calibrated to each individual. Two people might both have symptoms of headaches and fever, but if one was melancholic and the other sanguine, it was no good giving them both the same remedy. Even when physicians had diagnosed patients by letter in the past, they'd done so from detailed case histories, tweaking their treatments to account for the sufferer's diet and constitution. By contrast, mail-order medicine assumed that the same substance which cured a cold in Peter would do likewise for Paul.[15]

For medical practitioners, the impact was more complex, for this was the death blow to the last, limping attempts of the College of Physicians to regulate the medicine trade. First, physicians had competed only with their colleagues, of whom there were few, and sometimes with providers of household physic. Then, they found themselves jostling for power with an increasingly uppity body of apothecaries, surgeons and midwives. Now, they shared the marketplace with every Tom, Dick and Harry who cared to place a notice in the papers. However, if there was more competition for a slice of the medical pie, there was also a bigger pie to share. The medical elite carried on as they always had, but country doctors whose practice had been limited by time and transport could now minister to clients from Boscastle to Bradford without leaving the comfort of home, trading high fees for high volumes.

***

By the late seventeenth century, urban consumers could pick and choose from a smorgasbord of pharmaceuticals from different sources. But then as now, not all medicine came as pills and potions, nor were all such substances medicinal. Renaissance people had always been image-conscious. Beauty was a marketable commodity. The ideal for women was blonde hair, white teeth, small breasts and milk-pale skin like that of the young

Queen Elizabeth, or later, Charles II's famous mistress Nell Gwynn; for men, it was slim, shapely legs, small waists and aquiline features. As much as comeliness opened doors, however, bodily difference – scarring, disfigurement, limb loss – slammed them shut. Historian Simon Dickie has shown how eighteenth-century nobles revelled in organising ugly contests, making old women race one another for entertainment or staging marriages between 'deformed' people.[16] Little wonder then that as London was rebuilt after the Great Fire, image-conscious citizens looked to scaffold and plaster their own frontages. For many years, books of medicine had included remedies designed to improve the body rather than repair it. Gervase Markham's *The English House-wife* put receipts for pimpled faces alongside those for epilepsy and haemorrhage (take the white of ten hard-boiled eggs, crush them with copper sulphate and apply to the face morning and evening).[17] Mrs Meade's private recipe book had many medical receipts for a child named Nathanaell but also included a tinted lip balm made from butter, beeswax, raisins and root of alkanet (also known as dyer's bugloss).[18] In the eighteenth century, poet and professional misanthrope Jonathan Swift characterised women's dressing-rooms as hotbeds of filth and deception, in which eyebrows were supplied by mouse hide, full cheeks by cloth 'plumpers' kept in the mouth and a bright complexion by chemical 'daubs'.[19]

Nestled somewhere in between cosmetics and medicines were those aids that enhanced both form and function: false teeth and prosthetic limbs. In 1685, Charles Allen published the first text in English devoted entirely to dentistry: *The Operator for the Teeth*. Allen called himself a 'Professor' but stopped short of naming himself as a physician or surgeon. It is possible that he too was a 'quack' practitioner of the better sort, having spotted a cavity in the market. Dental hygiene at this point left much to be desired. Some people chewed mint or rubbed their teeth with sticks and cloths, but it was more to remove visible scraps and freshen the breath than to prevent decay. Meanwhile, the quantities of sugar imported into England tripled between 1560 and 1620, and its price halved during the seventeenth century, leaving more people than ever with a mouth full of rotten teeth. Allen had few suggestions to prevent disease, other than washing the mouth with rosewater and vinegar. There were even fewer cures (the earliest reference to lead fillings in England comes in 1715).[20] However, he reassured patients that when their teeth inevitably required extraction, 'we are not yet to despair, and esteem ourselves *toothless* for all the rest of our life'.[21] New teeth could be got from the mouths of either humans (dead or alive) or, with rather more hazard, animals such as baboons. Once extracted, they could be set into a mould to function as dentures, or simply shoved into the hole left by the missing tooth in hopes they would graft themselves in.*

Allen's keenness to insert the teeth of apes, dogs, sheep, human corpses or misfortunate servants into the mouths of his patients was particularly surprising given that by this point, there was a well-established tradition of false teeth made from bone. In 1648, the poet and Anglican cleric Robert Herrick wrote a biting verse about his enemy, 'Glasco', who

---

* Mercifully, there's no evidence Allen ever succeeded in finding any takers for animal teeth.

…had none, but now some teeth has got,
Which though they fur, will neither ache, or rot.
Six teeth he has, whereof twice two are known
Made of a haft [knife handle], that was a mutton-bone.
Which not for use, but merely for the sight,
He wears all day, and draws those teeth at night.[22]

As Herrick's poem showed, impetus for permanently inserted teeth rather than dentures may have come from social pressure; despite routinely wearing wigs, dentures and cosmetics, people did not like to have attention drawn to their artifice. Samuel Pepys noted crossly that 'Sir William Varren doth rail still against Mr. Turner and his wife (telling me he is a false fellow, and his wife a false woman and hath rotten teeth and false, set in with wire) and as I know they are so, I am glad he finds it so.'[23] In newspaper advertisements, dentists repeatedly stressed that their teeth 'could not be distinguished' from the real thing (perhaps true) and could be left in for seven years at a time (definitely not).[24]

The final prop holding up England's rotten, bashed-about, beleaguered bodies was prosthesis-making. The vast majority of people missing a limb used either crutches, a wooden peg-leg or a hook hand. In 1653, Ely Hospital placed an order for 144 crutches, costing £5 2s. The hospital carpenter William Bradley spent much of his time making wooden limbs: a wooden hand for an injured soldier named Fisher, a pair of legs for Thomas Swain, complete with straps and buckles.[25] These basic models were light and cheap and useful (sometimes in odd ways; in 1645, Parliamentary forces intercepted a Royalist soldier whom they suspected of carrying messages between regiments. A 'strict' search revealed nothing, until 'they began to examine his wooden leg which being taken off in the hollow thereof…letters were discovered').[26] But ever since Ambroise Paré had dreamed up complex mechanical limbs back in the sixteenth century, wealthy amputees had driven a niche market in sophisticated prosthetic technologies. At Ely, George Matheson received 'a new artificial leg with a leather box, plated all over with iron, complete with swivels and pins' for £3s 3s 11d.[27] It was a fair price, but you could have bought a hundred crutches with the same sum. On the continent, there were skilled men diverted from the clockmaking and jewellery industries who would create you an arm and hand with which to lift the hat from your head, a leg to ride on horseback, or enamelled false eyes and noses indistinguishable from the real thing.

By the eighteenth century, prosthetics had long escaped the control of physicians and surgeons, and the market in replacement parts was so hot that rival makers sought any means to run one another out of business. In 1710, a prosthesis-maker named John Sewers paid John Cluer of Cheapside to print him an advertisement. He'd recently moved from Cow-Lane near Smithfield to a more central location near Leather-seller's Hall in St Helen's (close to where the Gherkin now stands). However, an enterprising rival had taken advantage of the move to spread a rumour that he was dead. Sewers protested that not only was he still alive, 'He still maketh all sorts of ARTIFICIAL LEGS, for all Conditions, though never so difficult.' His legs, he boasted, could be worn even while an amputee's wound was still 'oozing', allowing them to 'walk with the greatest ease and safety, and take as firm and large steps as ever'. What was more, they weighed just a few pounds and were 'hardly to be discerned' by the casual observer.[28]

Cosmetics and prosthetics might have seemed far removed from each other, but it was no coincidence that they both evolved from medicine and began with medical practitioners, for like physic and surgery, they sought to repair and transform the body. They were all in the business of working wonders. But what happened when wonders became medicine?

'The Stroker': Valentine Greatrakes

In spring 1666, a meeting took place in the Palace of Whitehall. The palace was a rabbit warren of gilded rooms and winding corridors, seemingly far removed from the plague-ravaged city, and somewhere near the centre, Valentine Greatrakes – 'the Stroker' – was bowing deeply before Charles II. The two men regarded each other. One was an ex-Cromwellian, a sober man with a small mouth and heavy eyebrows, yet a curiously 'majestical' air.[29] The other was lavishly dressed and sporting a long wig, beneath which were large, protuberant features. He had quite recently had the corpse of Cromwell dug up and hanged, its head lopped off to be stuck on a spike atop Westminster Hall and its trunk tossed into a mass grave at Tyburn.

This meeting marked the apex of an extraordinary six months in which Greatrakes sailed to England from his native Ireland, cured (or attempted to cure) a myriad of diseases using nothing more than his innate healing powers, convinced some of the greatest scientific minds of the era of his abilities and promptly vanished home again. Greatrakes was what we might now call a faith healer. His nickname 'the Stroker' was attained because he treated people by gently stroking the area of their body afflicted by pain or disease. Stroke by stroke, he coaxed the malady outwards to their extremities until it could finally 'leave'

Valentine Greatrakes, 'the Stroker'. Line engraving by W. Faithorne, 1666. The Wellcome Collection.

via the fingers, toes, eyes, ears, nose or mouth (any attempts to expel disease via other orifices went unrecorded). How it worked, even he could not say, other than to murmur that he operated by God's grace. But in the bitter exchanges between the Stroker's supporters and his enemies, the lines between healing, showmanship, faith and suggestion were blurred as never before.

Only a few months before Greatrakes became the talk of England, he had been living a quiet, petty-gentry life in Affane, a small village in County Waterford. True, it had not always been such a placid existence, for Ireland was a hotbed of political strife and military activity. Greatrakes had served in the Cromwellian army there as a young man, then found himself temporarily relieved of his job as a justice of the peace when monarchy was restored in 1660. But nothing about this modest career suggested that, in about 1662, Greatrakes would suddenly find himself seized by a 'strange perswasion' that he possessed a 'gift' to heal sufferers of scrofula, a chronic, seemingly causeless condition which caused dramatic swelling of the lymph nodes.[30] His claim was especially audacious not because it had never been made before, but because it *had*. Scrofula was popularly known as 'King's-evil' because it had long been said that this disease could be cured by the touch of the monarch, and Charles II would often go about laying hands on his afflicted subjects – the seventeenth-century version of a politician kissing babies. To claim that a mere citizen could do the same thing was audacious at best, and at worst, treasonous.

As luck would have it, he soon found an opportunity to test his belief. Only a few days after Greatrakes's revelation, William Maher of nearby Salterbridge brought his son to Affane, hoping that Greatrakes's wife Ruth could treat him using what Greatrakes rather uncharitably described as her 'small skill in chirurgery'.[31] The boy was afflicted with swollen bluish lumps the size of a hen's egg around his eyes, face and throat, some of which threatened to rupture and create open sores. Ruth would not have known that the disease was caused by *Mycobacterium tuberculosis*, but she was well aware that it was difficult, often impossible, to cure. King's-evil killed many of its sufferers by creating a breeding ground for other infections, and so disfigured others that they could not venture outdoors without provoking horror and hostility. Greatrakes, on the other hand, was delighted. Here, as if divinely provided for him, was an opportunity to show his sceptical wife 'whether this were a bare fancy or imagination as she thought it, or the dictates of God's spirit on my heart' (one suspects from Ruth's reaction that Greatrakes may have had form for flights of the 'imagination'). He laid hands on the child's face, prayed in Jesus' name for him to be cured and instructed his father to bring him back in two days. Sure enough, when the boy returned, the lumps were 'strangely amended', though not quite gone. Within a month, the child was 'perfectly healed'. Greatrakes's new career had begun.

From the first healing, things escalated quickly. King's-evil baffled most physicians – even now, it requires many months of treatment with powerful antibiotics to cure – so sufferers from the surrounding counties flocked to Affane in hope of a miracle. According to Greatrakes, many of them had their wish granted. He claimed that he could heal virtually every case barring those in which the bones were 'infected and eaten', and soon, he experienced another 'impulse' telling him that he could also cure fevers, or 'agues'.[32] The stables, barns and malthouse on his farm were all turned over to temporary lodgings for the sick, as was much of Greatrakes's time.[33] At this point, the clerical authorities in Dublin were so aggrieved they ordered him to desist, but he took little notice. By 1665, he was ministering to all sorts of skin conditions, tumours, ulcers, aches and pains, claiming that people who could hardly crawl into his presence left it with a spring in their step. Patients were even crossing the sea from England, including the future Astronomer Royal John Flamsteed.

Over the water, one household received accounts of Greatrakes's cures with special interest. Lady Anne Conway, friend to Henry More and patron to Francis van Helmont, suffered from migraines that baffled the nation's best physicians. She was also an intellectual, learned in science yet open-minded in spiritual matters and happy to open her home to interesting visitors. She and her husband reached out to Greatrakes through their aristocratic friends in Ireland, and it was finally arranged that the healer would travel to meet Conway and see if he could help her at last. Unusually, Greatrakes agreed a substantial fee for his assistance. For other healings, he refused to take payment, but the Conways had money to burn, and he needed to cover legal expenses from a landowning dispute back in Ireland.

Whatever the Conways paid Greatrakes, it was too much. Within a fortnight, it became clear that stroking had no effect on Lady Conway's headaches. She would never find a reliable cure, and died thirteen years later aged just forty-seven. Yet, the Conways did not hold Greatrakes's failure against him; after all, he was only following in the footsteps of many other esteemed physicians. Instead, they invited him to stay longer at Ragley Hall, where over the next few weeks, hundreds of people from the surrounding counties came to be healed. As Greatrakes later recalled, 'many were cured of their diseases and distempers, and many were not'.[34] However, the inconsistency of his results did not seem to matter, for, with his growing reputation helped along by the Conways' many connections, Greatrakes was soon invited to nearby Worcester, where crowds gathered so thickly that he feared, 'I was like to be bruised to death'. From here, he received the order – the royal court did not make requests – to come to London, to Whitehall. He was about to meet the king.

In a 'brief account' of his life and cures written later that year, Greatrakes played down his time at Whitehall. He had, he said, planned to stay only three or four days before returning to Lord and Lady Conway, but was persuaded to extend his time in London owing to the many people there wanting cures.[35] In truth, his brush with royalty was a flop. Greatrakes performed his strokings in front of the king and other nobles, and some of them were successful, but he failed to cure the courtier Sir John Denham of his madness (some argued he'd made it worse), and as a result, Charles soon lost interest in this upstart. Once again, however, the Stroker's inability to cure the great and good made remarkably little difference to his popularity among ordinary people. He took up residence in Lincoln's Inn Fields, laying hands on hundreds of clients and gathering piles of testimonials to his work. Twenty-eight-year-old Anne Field, the wife of a pastry cook, suffered from violent headaches and partial blindness, and reported that in a few sessions of stroking she 'was perfectly freed of her pain in her head, and her eyes so amended, that she can now read a small print'.[36] Walter Dolle, the teenage son of a goldbeater, was completely paralysed on his right side, 'insomuch that he had no more motion on that side than a dead carcass'. He'd been thus for ten months, to the bafflement of physicians, but in 'five or six' encounters with Greatrakes, he regained his ability to walk, and was so recovered that he resumed his apprenticeship as an engraver.[37] Soon, it was said that even indirect contact with Greatrakes could work miracles. Elenor Dickinson, a Clerkenwell widow, hoped that stroking would relieve her tympany (accumulation of air in the abdomen), but found him surrounded by clamorous would-be patients. 'Not being able to come near him by reason of the throng', she reported that instead 'she snatched some of his urine and drank it, some of which she also put into her ears'.* Soon after, 'she voided near upon 4 gallons of water, with a great quantity of wind at her privy parts; and her belly which was before 2 yards in compass, doth not now exceed 3 quarters of a yard'.[38] Others swore that they'd been cured simply by being stroked with one of Greatrakes's old gloves.[39] Altogether, historian Peter Elmer estimates that over two hundred people testified to having been cured by Greatrakes in this period.

The Stroker's star was in the ascendant, but it was not all plain sailing. Greatrakes's claims offended churchmen and physicians alike. Even as Londoners were still flocking to his healings, a sceptic named David Lloyd published *Wonders no Miracles*, in which he made the case that Greatrakes was no more than another itinerant quack, a man who despite his lofty spiritual claims 'doth not look as if he were much troubled with *impulses*'.[40] Tales of his success were, Lloyd argued, merely sleight of hand on a grand scale; Greatrakes chose his clients not through divine inspiration as he claimed, but according to who looked most

* Sadly, there's no mention of how she got Greatrakes's urine. Perhaps an enterprising hanger-on was selling it by the jar.

likely to recover. He told them to do quite obvious things such as washing their sore eyes with water, and then claimed the credit when they recovered naturally days, weeks or even months later. Lloyd continued:

> That the giddy multitude should follow any strange thing…is no wonder, especially in such a year of expectation as this is: But they follow him not in any place so eagerly at first, as they leave him discontentedly at last: he is not so much cried up in the places where he comes, as he is cried down in the places where he hath been: and he removes from place to place, not so much to communicate his virtue, as to save himself, being not known two nights together in one lodging.[41]

These were the age-old complaints about itinerant and unlicensed healers: they lacked morals; they moved around to outrun the bad reputations they accrued; despite the hype, those unlucky enough actually to meet them soon realised they had been duped. Lloyd gathered his own body of evidence to counter Greatrakes's supportive testimonies: a maid 'cured' by the Stroker of mutism, but still only able to say one word; a gentleman whom he'd determinedly stroked only to end up causing an aposthume (pustulent tumour); a gentlewoman whom 'he went to dispossess of a devil, when she was only troubled with the wind'.[42]

Damning though it was, Lloyd's critique also contains hints as to why Greatrakes became so phenomenally popular almost overnight. In 1666, the 'year of expectation', people whose lives had been destroyed by fire and plague were disinclined to listen to the chidings of physicians who had proved entirely unequal to these apocalyptic times. Greatrakes's methods might have seemed outlandish, but they offered something different at just the moment when many clever and well-respected men had tired of Galenic physic and were looking for a new mast to which they could nail their colours. Henry Stubbe, a learned physician recently returned from Jamaica, published a fifty-page booklet speculating about the possible scientific basis for 'Greatarick's seemingly miraculous cures'.[43] Greatrakes himself could call on testimonials from many respected intellectuals: Lady Conway's old friend Henry More; the Bishops of Chester and Hereford; the London magistrate Sir Edmund Berry Godfrey; the poet Andrew Marvell; and the eminent scientist and mathematician Robert Boyle.[44] Some of these men neither knew nor particularly cared whether Greatrakes's stroking was really effective, but they admired his modesty and morality – especially when contrasted with 'merry monarch' Charles II, a man described even by his friends as 'roll[ing] about from whore to whore'.[45]

In the end, Greatrakes's star burned brightly and briefly. Later that year, he left London and returned to Ireland, for reasons he never explained. Perhaps he grew tired of the attention and adulation, or could not tolerate London's dirt and noise after the green hills of Affane. Or maybe, in the wake of Lloyd's treatise against him, he thought it best to leave town before he lost his lustre. Public opinion could turn on a sixpence, and a popular ballad told gleefully of how a plucky farmer named 'downright Dick' beat a boastful mountebank with such ferocity that he 'did crack his crown…And the blood did trickle down / In more than twenty places'.[46] Greatrakes briefly returned to the capital in 1668, but he slipped in and out with none of the fanfare of his previous visits. For the most part, he lived quietly, farming his estate and treating only those who came to his door, until his death in 1683.

***

Medicine was never *just* medical. Valentine Greatrakes's rise had as much to do with people's yearning for a new, more peacable kind of politics as with their belief in his powers. Healing men and women, rich and poor, confirmed Royalists and others longing for the return of the commonwealth, the Stroker became symbolic – at least to his supporters – of a future in which Christian rectitude won out over self-interest. What was more, all those involved with medicine at this time – the authorised, the unauthorised and their patients – were probing the boundaries of this ever-growing field, testing out what belonged to the healing sphere. We can think of the people in this chapter as placed on a Venn diagram with overlapping circles of commerce, faith and science, where the middle is very large indeed.

As for the quacks, did they really believe in their own cures? As we have seen, this group was more varied than their critics would give them credit for. There were certainly villains among them, people who cared nothing if their cure-all potion were completely ineffective, eager to use a suffering patient to test their prowess with a lancet, and ready to take money from desperate parents knowing full well that their skills were smoke and mirrors and of no use to an ailing child. Yet there were others with genuine aptitude for their craft. After all, the business plan of a true fraud was limited; you could pull the same trick only so many times before being run out of town or worse. Then there were the many and varied specialists – prosthesis-makers, tooth-drawers, cosmeticians – whose only crime was to excel in fields which hadn't been formally invented yet. That history has often tarred all these people with the brush of 'quackery' says less about their skills, and more about the way that the learned and powerful sought to control both the medical market and its legacy.

The Hospital of Bethlem [Bedlam] at Moorfields, London: seen from the north. Engraving. The Wellcome Collection.

CHAPTER EIGHT

Care-Takers and Criminals: Hospitals and the Welfare State

On the last Saturday in January 1722, in the heart of the City of London, well-to-do residents are enjoying the conviviality of taverns and coffee-houses, taking refuge from the biting cold. Newspapers are passed between patrons and read aloud, interesting morsels chewed over and meatier issues loudly debated. It takes a lot to surprise these men and women, whose lives play out between the noise and smells of Smithfield Market, the miasma of the river and the stolid grey and white bulk of St Bartholomew's Church. On this day, however, the *London Journal* contains something rather remarkable. It reads:

> On Monday last part of the right Leg of a man was found in a cellar window in Bartholomew Close, which probably may have belonged to some patient in the neighbouring hospital, that has undergone an amputation; some will have it otherwise, and to be a limb of one that has been murdered. If the owner be not living, the flesh on it, shewed plainly that he has not been long dead.[1]

There are many lost and found advertisements in the *Journal.* Hats, horses, boys, servants, slaves, falcons and wives may all go astray, but to misplace a leg is rare indeed, and the newspaper's laconic tones do not do justice to the strangeness of the phenomenon. There are practical matters at hand. Where is the previous owner of this wandering limb, and does he or she know that their leg has gone walkabout? If amputated, shouldn't it have been given a Christian burial? Or, has the body of the owner fallen victim to the incipient trade in corpses for anatomisation, the leg being surplus to requirements?

Swirling under the surface are more abstract questions. What happens to a person if, on the day of resurrection, the body which arises from the grave is missing some of its parts? If the stray leg had been eaten by one of the pigs that roam the streets, or tumbled into the river to be nibbled by fish, would the eater of that pig or those fish have been a cannibal? These are puzzles which occupied John Donne as he preached in nearby St Paul's Cathedral a century prior – they still trouble philosophically inclined Christians.

All the questions, pragmatic and metaphysical, go unanswered. Over the next six months, the *London Journal* mentions men stabbed in the leg, trusses for misshapen legs, salves for rheumatic legs and numerous goings-on at the Leg pub. But the strange sight of the leg in the window is never spoken of again.

* * *

Today, St Bartholomew's Hospital – Bart's, as it is known by Londoners – is a centre for cardiac and cancer care, redeveloped in 2016 to the tune of £500 million. Nevertheless, it is one of the world's oldest hospitals, still occupying the site where it was first founded in 1123 during the reign of Henry I. The *Civitas Londinum* map of London printed in 1561 shows the medieval hospital chapel overlooking 'Schmyt Fyeld' and its roughly sketched but clearly rather lively inhabitants: two men galloping on horseback, one wielding a stick; two figures clasped in an embrace, or maybe locked in combat; one man seemingly hanging on to another to stay upright; and somebody minding their own business, tending their cow. The oldest parts of the hospital survived the Civil Wars, the Great Fire of London and the Blitz.

But in 1722, Bart's was the sort of place where one could lose a leg twice; first, by having it amputated, and secondly, by mislaying the disembodied part. Was the curious incident of the lost leg a simple mishap, or an indication that hospitals in this period were dysfunctional, even dangerous – a place of harm rather than healing?

* * *

Health began at home in the Renaissance. But if you had no home or were so poor that home could not supply your basic needs, the outlook was grim. Back in the medieval period, abbeys and monasteries had helped care for the sick, supplying premises, staff and cash.* London's two main hospitals, St Bartholomew's and St Thomas's, were both founded in the twelfth century by Augustinian clerics: the former bankrolled by a courtier-cum-monk named 'Rahere', the latter established as a part of St Mary Overie priory, but later renamed for the murdered Archbishop Thomas Becket. Even these most prominent hospitals were

* 'Hospital' in this period was a broad term. Anywhere that supplied hospice could be deemed a hospital, from plague houses and leper colonies to workhouses and alms-houses.

tiny. When St Thomas's was rebuilt in 1212 after a fire, it was expanded to forty beds, and Bart's cannot have been much bigger, squashed into one of the most crowded areas of London. Even for the city as it was then, with about fifty thousand inhabitants, this was hopelessly inadequate. But it was not only the hospitals that provided care. Every social safety net in sixteenth-century England was built on the foundation of the Catholic Church, from alms-houses to leper colonies. Such activities were funded by wealthy patrons who bought prayers for their own souls and those of their deceased loved ones, and by poorer benefactors seeking to bolster their own spiritual credit with alms-giving.

Thus when Henry VIII broke with Rome, it boded catastrophe for the health of the very poorest. Technically, hospitals were separate from monasteries, but in the feeding frenzy of the Dissolution, these institutions – peopled by clergy and often built in the same style as monastic buildings – proved easy pickings. In London, Bart's and Thomas's shut down, as did St Giles's in Holborn. The story was repeated around the country: the great St Leonard's in York, with its vaulted stone basement; St John the Baptist's in Bristol, whose seal bore the emblem of the lamb of God; Burton Lazars in Leicestershire, where Henry's own father had been a patron; the Maison Dieu in Dover, afterwards turned into a tavern; and the leper house in Bury St Edmunds. When the purge began in 1536, Henry's stated aim was to reform monastic life, not obliterate it. The clean-up started with all institutions worth less than £200, which included most of the smaller hospitals and alms-houses where, according to Henry's right-hand man Thomas Cromwell, the worst improprieties were to be found. But in the midst of an expensive war with France, the new source of income proved intoxicating. Virtually every abbey was dissolved, and even those hospitals which were spared struggled to continue when their source of income, provisions and staff was abruptly cut off.[2]

Henry's advisors specialised in helping him make messes and then cleaning them up. In the years after the Dissolution, the king gifted the sites of St Bart's and Bethlem Hospitals (though not the funds to actually run them) to the Corporation of the City of London, a civic organisation comprised of wealthy aldermen and merchants. Bart's promptly installed Thomas Vicary, the shrewd petitioner who'd successfully lobbied the king to establish a Company of Barber-Surgeons only eight years previously. The hospital would later employ John Woodall, the ship's-surgeon who'd counselled his readers on how to prepare for a battle at sea. Henry's son Edward VI did more, refounding St Thomas's and establishing two new hospitals: Christ's, for poor orphan children, and Bridewell, a workhouse for the so-called 'idle poor'. Successive monarchs were willing enough to grant permission for new hospitals provided they did not have to fund them; Elizabeth's favourite courtier Robert Dudley successfully applied for permission to establish a hospital for retired soldiers in Warwick in 1571. Through the efforts of individuals and civic societies, by piecemeal charity and patronage, the system limped along. But by the end of the sixteenth century, nobody could deny that it was ready to collapse.

* * *

For all the courtly pomp and graces of Queen Elizabeth and her court, the country over which she presided was full of disabled, sick and starving people. Crops repeatedly failed, leading to rampant inflation. Plague periodically swept through towns and cities. And swathes of the male population lost their health, if not their lives, in the army and navy. Every town and village had men who left as young sparks and returned limping, blind, mad or maimed.

The homecoming welcome they received was mixed to say the least. On one hand, a martial injury – say, a dashing scar or a limp – could be proof of one's valiant service. In Shakespeare's *Henry IV, Part Two*, the lovable ne'er-do-well Falstaff talks of how he will pass off the effects of venereal disease and overindulgence as hurts sustained in military action:

> A pox of this gout! or, a gout of this pox! For the one or the other plays the rogue with my great toe. 'Tis no matter if I do halt [limp]; I have the wars for my colour, and my pension shall seem the more reasonable. A good wit will make use of anything: I will turn diseases to commodity.[3]

Meanwhile, plays like *The Shoemakers' Holiday* portrayed plucky former soldiers embraced by their communities (and their sweethearts) despite their physical flaws. There was even a rumour put about that men who'd lost a limb would be particularly virile in the bedroom, since the blood that used to circulate to that part of the body had to be accommodated…elsewhere.

For all this jolly posturing, the reality of life for many disabled people was grim. If they could not return to their previous trade – and manual labour was by far the main sort of employment – they might turn to begging, and in doing so, risk their lives all over again. In 1572, the Vagabond Act was brought in, obliging parishes to care for 'impotent' people who had been resident for at least three years. But those same laws allowed corporal punishment for adults found begging, and while soldiers and sailors were supposedly exempt (along with victims of highway robbery), they had to produce a licence from two justices to avoid being run out of town. If disabled people thought that the obvious limitations of missing a limb or an eye would protect them from such hostility, they were sadly mistaken, because the fear and loathing of vagrants reached such a pitch that no proof was strong enough to secure a person from accusations of 'idleness'. Back in the sixteenth century, Ambroise Paré – in so many ways a compassionate and progressive man – nonetheless recalled with delight how he and his brother had uncovered the impostures of a woman beggar claiming to suffer from breast cancer. Noticing that her face was plump and rosy despite the hideous sores on her body, he demanded she be taken away and strip-searched. In her armpit the magistrate found

> a sponge soaked and imbued with animal blood and with milk mixed together, and a little elder pipe through which this mixture was conducted through fake holes in her ulcerated canker…Then he took some warm water and fomented [soaked] the breast, and having moistened it, he lifted off several black, green, and yellowish frog-skins, placed one on top of the other, stuck together with bolearmenie [bole armenic, a kind of clay] and egg white and flour.

The beggar woman confessed that she'd been put up to the deception by her partner, who usually pretended to have a terrible leg ulcer (actually an ox's spleen tied on and pierced in several places). However, the magistrate was unsympathetic: 'he condemned the slut to get the whip, and [she was] banished out of the country, but not before being curried with lashes of a whip made of knotted cords, as one did at that time.'[4]

In England, too, people feared hunger and the spread of plague, and they looked for reasons not to trust strangers who appeared asking for alms. They found many, for the era's best playwrights delighted in the theme of the fake beggar. In 1608–9, Thomas Dekker, a prolific playwright and writer-for-hire, released a series of pamphlets detailing in colourful terms the supposed skulduggery of confidence tricksters: 'clapperdudgeons', 'mawnders', 'palliards', 'praters', 'frigges', 'swadders', 'curtals', 'swigmen', 'upright-men' and 'counterfeits'.* Those who claimed to be soldiers, he wrote, always travelled in twos or threes, carrying cudgels, 'knocking at men's doors as if they had serious business there'.[5] Often they were covered in gruesome-looking wounds, but it was all a ruse, for they'd created the bloody sores themselves, taking unslaked lime, soap and rust, spreading it thickly on a piece of leather and binding it hard to the limb until it blistered, then applying a linen cloth which, when removed, would take the skin with it. 'Thus do you see how our mawndering counterfeit soldiers come maimed.' There was no national crisis, he insisted, except an epidemic of scroungers, who in private peeled off their sores, unbound 'stump' legs bent double into trousers, opened blinded eyes and danced a merry jig as, in their thieves' cant, they discoursed of the day's frauds.[6]

---

* Dekker was also the author of *The Shoemakers' Holiday*, so he played both sides of the debate, portraying a wounded soldier sympathetically on one hand, and accusing all such men of deception on the other.

It was a convenient fiction, but an unnecessary one. People were perfectly capable of mistrusting people with genuine disabilities. Opinion varied over why people with disabilities ought to be feared, whether it was because their difference showed that they'd incurred God's displeasure, or, as the philosopher Francis Bacon proposed, the hardships of their lives made them malicious.

In 1614, a pamphlet describing some of the Old Bailey's most scandalous trials told the story of John Arthur, a young man born 'lame and limbless'. Arthur 'lived and maintained himself with the charity and devotions of alms-giving people', but instead of being grateful, spent the donations he received on getting drunk, 'which is evermore the receipt of such begging vagabonds'.[7] However, this was not what landed him in Newgate jail. Arthur had begun a relationship with an unnamed young woman, also disabled, and the court heard how when he tired of his companion,

> The cripple…took the woman's own girdle, and putting the same slyly about her neck, where though nature had denied him strength and limbs, yet by the help of the devil, which always adds force to villainy, he made means in her sleep to strangle her.[8]

One might have thought that in such a case, the defendant's obvious disability might go in his favour. It is not clear whether Arthur had any arms or legs, but he was acknowledged to be 'not able of his own strength to help himself', so managing to strangle his partner was surprising to say the least. Yet the discrepancy was explained away as the help of Satan, who 'will still make…

those that be the most weakest, to be of the vilest thoughts'. The account concluded with a lengthy and poorly written poem, allegedly by Arthur, in which he explained that he was a lifelong wrongdoer, and when he was enraged by his partner's demand that he 'make her married wife', the devil lent him strength to overpower her. For doing so, he was almost certainly executed at Tyburn's notorious 'hanging tree'.

The story of Arthur's life and death was told with a savage delight that reflected the fractious mood of the times. At the turn of the century, following a series of failed harvests, Elizabeth's government had instituted a flurry of new laws designed to staunch the flow of vagrants into towns and cities. In 1597, a Poor Law decreed that each parish ought to have an overseer of the poor who would collect poor rates from property owners, distribute food to the needy and oversee the parish poor house, if they had one. In 1601, this law was updated to become the first national system, levying a poor rate on every parish which would be used to sustain the so-called 'deserving poor', that is, children, the disabled and the elderly. At the same time, the *Act for Punishment of Rogues, Vagabonds, and Sturdy Beggars* ordered that any able-bodied man or woman found destitute outside their own parish should be whipped and sent home, where they would be confined to the local workhouse. The law was applied with vigour; records supplied to the Privy Council by various town authorities show that in the 1630s, 132 vagrants and vagabonds were punished and expelled from Hertfordshire, 296 from Devon and Cornwall, 616 from Kent, Sussex and Surrey and an impressive 916 from Lancashire and Westmorland.[9] They included maimed soldiers and homeless people begging for alms, but also fortune-tellers, pedlars, minstrels, gypsies, seasonal labourers, conjurors, criminals and those habitual troublemakers, Morris dancers.[10]

* * *

In the Elizabethan carrot-and-stick approach to the problem of public welfare, the carrots were meagre and the stick heavy. Yet there were some attempts to provide for the poor and sickly. A third act of 1597, the *Act for Erecting of Hospitals*, encouraged philanthropists to set up hospitals and alms-houses, and the number of such institutions grew slowly over the coming decades.[11] Often they were far from what we would recognise as a hospital today. Some were essentially retirement villages, providing accommodation, food and some care to people who were not ill, but simply too old to work and without family who could be forced to take them in. Orphanages were likewise categorised as 'hospitals' – Christ's Hospital, near St Paul's Cathedral, was one of them. At the other end of the scale, lazar-houses and pesthouses were populated by the leprous and plague-ridden. Their purpose was less to cure their patients than to keep them from infecting other townsfolk, though as leprosy subsided, the lazar-houses were often repurposed for treating venereal disease. In the middle were places like Sutton's Hospital, established by the low-born but fabulously wealthy merchant Thomas Sutton in 1611. Occupying the old 'Charterhouse' in Islington, it provided a home for eighty ex-servicemen and a school for boys. However, it also acquired a far-reaching reputation for medical expertise, so that in 1627, a Lancashire wool-maker named Robert Butterworth appealed to his local court to be given a certificate that would allow him to travel to London unmolested to seek treatment for his facial cancer:

> Being a very poor man and charged with a wife and one child [he] is troubled with a grievous disease called the canker which begun in his face about two months ago, and hath ever since grown worse and worse...and your petitioner much doubting that he can get remedy for his said disease in the country intendeth to travel up to London in hope there to get the same cured – and for that your petitioner hath had conference with one of his neighbours a poor man who got the like disease cured at London being admitted into a hospital there called Sutton's Hospital.[12]

Butterworth's petition shows that despite their manifold failings, some hospitals, particularly in London, nevertheless became known as centres of excellence. Many also functioned more or less informally as teaching institutions, where 'cub' surgeons trailed the wards after their masters. Even senior figures could learn much from the volume and severity of the cases they encountered there, as when ship's-surgeon John Woodall was appointed surgeon for St Bart's in 1616.

From his years treating combat injuries, Woodall had a particular interest in gangrene, 'a disease woeful, painful, horrible, and fearful to mankind'.[13] When the disease reached its most advanced stage, 'sphacelus', the affected limb turned black and putrefied on the body. Without amputation of the part, the sufferer would soon die. In St Bart's were many such cases, and Woodall recalled one from 1617 in which gangrene had rotted the leg of a hospitalised man until he was 'as feeble and weak, as possible a living creature might be'.[14] The sufferer was so ill, Woodall was at first reluctant to treat him at all; believing surgery to be futile, he wrote, 'I intended the patient should die by nature, rather than to be killed by art.' Yet, he resolved that if he could remove the limb by cutting through the dead area rather than above it, the operation might be justifiable. He admitted it probably wouldn't succeed – the prevailing wisdom was to cut above the affected part – but at least the patient would not feel pain, since the gangrenous part was insensible, nor die from blood loss, since it was also bereft of circulation. To Woodall's astonishment, however, the day after the operation he found that his patient began to dramatically 'cheer up'.[15] They stayed in hospital for a further ten weeks, until 'being then sound and lusty', they 'gave thanks to the Governor of the hospital, in full payment of his cure, and so departed from the hospital upon a leg of wood'.[16]

Woodall was delighted with his new method, which did away with the agony of amputation, and he began to use it on almost all gangrenous patients. It is a mystery why it worked; now as then, surgeons advise that mortified limbs be removed or gangrenous parts debrided by cutting into live flesh, to avoid the rot spreading farther. Nevertheless, 'not above four of each twenty' of the operations he did at St Bart's resulted in death, a commendable success rate for the time. Woodall also sought to endear himself to readers by pointing out that though he had lately heard of an intriguing method from the East Indies, in which felons who'd had their feet amputated used large reeds or bamboo as prostheses, he had so far resisted the temptation to try this as an experiment on any of the hospital patients. Though he could easily have done so, he said, 'I have, I confess, ever been tender in that point, and loath to put any one patient against his own free will, upon new inventions, by beginning new practice upon him; every patient in such a case having just freedom to choose his way.'[17]

Woodall was proud of himself for declining to experiment on impoverished patients against their will. If his attitude reflects the times – and Woodall was a well-respected, well-connected member of the medical establishment – then the care provided by free or cheap hospitals came at the hidden cost of relinquishing all control over one's treatment. Woodall's boasts reveal the frightening power that a single prominent surgeon or physician could hold over a whole institution. But the most shocking example of autocracy in action comes from a quite different kind of hospital – the notorious, nefarious Bedlam.

The crooked Bedlamite: Helkiah Crooke

Bedlam – or to give it its proper name, Bethlem Hospital – was almost as ancient as London itself. It predated Hannah Allen's violent madness and Robert Burton's intractable melancholy, and was older than the College of Physicians, the Company of Barber-Surgeons and the Society of Apothecaries put together. In fact, Bedlam was one of the oldest hospitals in Europe, narrowly bested by its neighbour St Bart's. It was established in 1247 when a wealthy tradesman and former Sheriff of London named Simon FitzMary gifted a parcel of land just outside Bishopsgate to the Bishop of Bethlehem, Goffredo de Prefetti. There a small religious institution was to care for the sick, the elderly or those who simply had nowhere else to go, but Bethlem did not become associated with the mentally ill until over a hundred years later. Before 1377, 'distraught and lunatike' persons were housed in the Stone House in Charing Cross.[18] But according to the seventeenth-century writer John Stow, a medieval king – perhaps Richard II – took a dislike to having the mad housed so near his palace and ordered their removal to Bethlem, safely out of earshot.

Like many Renaissance hospitals, Bethlem was idiosyncratic from the start, in many ways as mad as those it housed. For one thing, its management structure was bizarre, first resting with the Church, then divided between the Church, the king and the Corporation of London, and then throwing off its religious association entirely with the Reformation. After Henry VIII's death, the Court of Aldermen took over, but soon decided that the day-to-day business of caring for the insane would be better handled by the nearby Bridewell Hospital, which was in fact less a place of healing than a prison for unruly women. Bethlem and Bridewell both answered to a board of governors consisting not of medical practitioners but of local politicians and upwardly mobile merchants, who bought their places by donating £50 to the Bethlem-Bridewell pot. Caught in the power struggle between town and Crown, the hospital was a political hot potato, and matters were made worse by the second of its idiosyncrasies – its fame. On its Bishopsgate site, Bethlem was crammed into a plot of two acres, which had to accommodate a chapel, staff accommodation and kitchens as well as forty cells. Yet it was not oversubscribed. There were plenty of mad in London, but in the sixteenth and early seventeenth centuries, they were still predominantly being cared for at home. 'Bedlam', as it was now known, had already become notorious.

The first use of 'Bedlam' to mean chaos and disorder was in 1627, when the poet Michael Drayton wrote that England was a 'mere Bedlam' in which 'We all lye raving, mad in every sin'.[19] For years before this, however, authors had depicted Bedlam itself as an almost mystical fount of misery and madness, an institution so all-encompassing that its patients too became known as 'Bedlams'. The most famous example comes from Shakespeare's 1608 *King Lear*, in which the faithful Edgar disguises himself as 'Poor Tom o'Bedlam' to help his beleaguered sovereign. Yet the audience at the Globe would have recognised Tom as a long-standing stock figure, doing the rounds in ballads where his mental illness blends with beggary, mysterious folk wisdom and a surreal sense of time out of joint:

> Of thirty bare years have I
> Twice twenty been enragèd,
> And of forty been three times fifteen
> In durance soundly cagèd*
> On the lordly lofts of Bedlam,
> With stubble soft and dainty,
> Brave bracelets strong, sweet whips ding-dong,
> With wholesome hunger plenty,
> And now I sing, Any food, any feeding,
> Feeding, drink, or clothing;
> Come dame or maid, be not afraid,
> Poor Tom will injure nothing.

* Obviously, it's not possible to be 'enraged' forty years out of thirty, or imprisoned forty-five years out of forty. Poor Tom's grasp on time is warped by madness and the rigours of the asylum.

The dramatic figure of the ragged, raving, beaten Bedlamite struggling across the stage was pitiful, but the truth was just as bad. On 4 December 1598, some of the hospital's governors inspected the institution. They found it freezing and filthy. The twenty inhabitants huddled round the only heat source in the building, the kitchen fire, and the whole place reeked with the stench of human waste from the two open sewers that skirted the site.[20] They also noted that several women had given birth in Bedlam over the last decade, which suggested that the unsegregated female patients were being abused by fellow inmates, staff or both. It was no wonder the hospital operated at only half its capacity – most patients would rather die on the streets than enter its 'care'.

Our picture of who these unfortunate inmates were is hazy, based largely on the records held by the nearby St Botolph's Church of inmates who were buried there. Men and women appear in almost equal numbers; the days of women being committed by tyrannical husbands to get them out of the way still lay ahead. They ranged widely in age – a fifteen-, sixteen- and eighteen-year-old were all among those buried, as were several patients over fifty.[21] While Bedlam's association with Bridewell meant that it received many vagrants and paupers, there were also residents there from comfortable backgrounds. The St Botolph's register lists a 'Lady Marye Bohun' among those who died at Bedlam, though it also gives her age at death as a rather optimistic 140 years. The one thing all these people had in common was that they were mad in ways that made them dangerous or disruptive to their families or communities, and they were believed to be capable of recovery (meaning that 'simple' or 'foolish' people with learning disabilities were excluded, at least in theory). But of the twenty inmates listed in 1598, only one had been admitted in the last year. At least six had been there for over eight years, and one had been confined for a quarter of a century.[22]

After the disastrous 1598 inspection, those nominally in charge of Bedlam were appalled by what they had seen, yet curiously unable to make meaningful changes. They tinkered around the edges, raising donations for more food and better habitations, but the real problem was that nobody in the hospital seemed capable or even *interested* in curing their patients, for the keepers of Bedlam were not doctors but jailers. Something needed to change, and once again, transformation came not through careful planning but in the form of one medical man on the make.

Helkiah Crooke was in medicine for money and power. This in itself was not unusual; as we have seen, the medical world was a good place for enterprising young men to make their fortunes. However, Crooke made avarice his guiding light. Born in 1576, he had come from a middling family in Suffolk, genteel enough to educate him, but impoverished enough that he needed a scholarship to study for his medical degree at Cambridge. Having attained the qualifications, he duly applied to the Royal College of Physicians for a licence to practise in London, but he was apparently so rude during his interview – including accusing a colleague of malpractice – that he was sent away and denied admittance for another two years.[23] It did not slow him down. Crooke was a gifted scholar, especially in anatomy, and in 1615, he wrote *Mikrokosmographia*, a gigantic work compiled and translated from various European texts, with some of Crooke's own opinions tossed in for good measure.[24] It was published by William Jaggard, himself a barber-surgeon's son, who had recently gone blind as a result of syphilis (one of his apprentices, Thomas Cotes, would go on to inherit the company and print Ambroise Paré's *Workes*). Though a lucrative bestseller, the book managed to attract the ire of both surgeons and physicians: the former, because Crooke was trampling on 'their' territory, and the latter because the book included scandalously large and detailed pictures of human genitalia. No matter. Whilst the college authorities were still sputtering their ire, Crooke had set his sights on a new project – managing the most famous hospital in London.

Crooke took over the running of Bedlam in 1618, after a concerted campaign to oust the previous keeper Thomas Jenner, whom he accused of 'misgoverning and misbehaving himself' while in post. His cause was helped by the fact that he had at various times been personal physician to James I; *Mikrokosmographia* was dedicated to James, who Crooke described breathlessly as the 'most literary' of kings. The appointment should have been a breath of fresh air for the gloomy old hospital, since this was the first trained medic to hold the post of keeper. But before long, people began to suspect that Crooke was living up to his name. Things at Bedlam got even worse. Having worked so hard to secure his post, the new keeper swiftly lost interest, and by 1625 was visiting the hospital only a few times a year, leaving the day-to-day management to his son-in-law Thomas Bedford.[25] Rather than attempting to actually cure patients, Crooke spent his energies on running his private practice and complaining to anyone who would listen that the hospital was underfunded. Bedlam was by now attracting visitors, who came to gawp at the lunatics confined within, but this acted as another income stream rather than a motivation to clean up the place.

In 1631, matters finally came to a head. Two of the hospital's governors made an inspection and discovered a situation even bleaker than that of 1598. There was hardly a scrap of food in the entire building, and some patients were 'likely to starve'.[26] A week previously, the entire patient body had been fed for the day on only 4lbs of cheese. A less self-assured man might have apologised for the squalor and tendered his resignation, but instead, Crooke decided to brazen it out, and insisted this was evidence of the need for more money to be siphoned into Bedlam. His effrontery could only work for so long, and by 1633, King Charles I had ordered an investigation. It was soon revealed that Crooke had been profiteering left, right and centre. For each patient admitted, he charged between fifteen and twenty shillings, which went straight into his own pockets.[27] He had also sold land belonging to the hospital and stolen charitable donations meant for the upkeep of his poorest residents. Altogether, the investigators judged that Crooke had been making over £100 each year from Bedlam, and it was the embezzlement, not the mistreatment of patients, which finally led to his dismissal.

Crooke's reign over Bedlam exemplified all the worst parts of Renaissance medical institutions: ineffectual and politically motivated governance; autocratic leadership; almost total lack of inspections or standards. Yet, it also yielded changes which would eventually impact all hospitals. Partly to cover his own laziness, Crooke had brought in new staff to Bedlam, assembling a skeleton which would be fleshed out in the coming years. He employed a steward to deal with paperwork and administration, a matron to oversee the staff and assist with female patients, and a visiting surgeon to bleed the inmates and treat their wounds. When Crooke was ousted, the governors wisely decided to separate the roles of keeper and physician. Even more sensibly, the Court of Aldermen, which had endured fourteen years of Crooke's complaints and several decades of headaches from Bedlam, opted to give up the hospital as a bad job, and relinquished all matters of management and staffing to the Bridewell governors.

As for Crooke, the revelation of his callousness, deception and outright criminality did not humble him, and he campaigned for reinstatement to his post for the next eight years, without success. He had become an embarrassment to the College of Physicians, who eventually paid him £5 to relinquish his membership in 1635. On 18 March 1648, he was buried in the parish of St James's in Clerkenwell, remembered by his friends as a flawed genius and his enemies as an unrepentant scoundrel.

* * *

By the middle of the seventeenth century, the hospital system was administratively labyrinthine, corrupt and wholly inadequate to the needs of those it served. It was also about to face its biggest challenge yet.

In August 1642, war broke out between Charles I's Cavaliers and Oliver Cromwell's Roundheads. It was the first war of the newspaper age, and in the battle for public support, hospitals and welfare were at the frontline. From the first sallies, partisan newspapers on each side reported – often with scant regard to the truth – the great losses their enemy had suffered and the way opposition troops were left unpaid, demoralised and unrewarded for their sacrifices. The Royalist *Mercurius Aulicus*, for instance, alleged that Parliamentarians abandoned maimed servicemen entirely, for 'the number of the rebels' wounded men swells up so high, and their money sinks down so fast…that they judge no soldiers fit to live longer than they are able to fight'.[28] *Mercurius Britannicus*, the principal Parliamentarian paper, swiftly retaliated by claiming that if Royalist soldiers were not deemed immediately curable, 'their brains are either beaten out with the butt-end of a caule [carpentry tool], or at the best, left languishing under the hands of some ignorant poultice-plotting-horseleech, that bears the name of a chirurgeon, the pupil of some country midwife'.[29] (Even in the midst of war, the squabbling over which medical practitioners were legitimate continued.)

As a sick or injured soldier, one's fate depended equally on affiliation and location. For Parliamentarians in or near their side's stronghold of London, treatment was relatively civilised. Several hospitals were repurposed; Christ's Hospital, for example, was ordered to give preference to the orphans of soldiers on the Cromwellian side, and to pay out pensions to widows and injured Roundhead servicemen. St Thomas's and Bart's, both already creaking under the weight of the civilian population, now took in extra patients, housing an average of fifty-six soldiers at any time.[30] The influx caused tension in Bart's not only because they were ill-provisioned for this increase, but because the institution had been traditionally Royalist. The king's loyal supporters were now obliged to care for men who wished for his death.

Specifically for military men, the Savoy Hospital (on London's Strand, on the site which would later be occupied by the Savoy Hotel) was re-founded after a decades-long hiatus. Like Bedlam, it had been mismanaged, with its Master Thomas Thurland found guilty in 1570 of stealing money from the hospital, having sex with the nurses and even selling off the beds.[31] Having gradually been given over to housing and workshops, it was now brought back into use with a staff consisting mainly of war widows. Each hospital was sorely needed, for soldiers were brought in with appalling wounds. Historian Charles Carlton describes how one Parliamentarian soldier spent six months at Bart's having sixty bone splinters removed from his shattered legs. He left the hospital alive but 'almost continually in intolerable pain', eventually being awarded a scant £4 annual pension for his trouble. Another private had a broken skull which was patched up with silver plate, as well as numerous injuries including a broken jaw, a 'maimed' hand and an abdominal wound.[32]

We know a good deal about the state of Parliamentarian hospitals because of the diligent record-keeping of Cromwellian leaders and their liberal use of the press to help them raise funds to keep the war machine running. In a printed pamphlet from 1645, the mounting costs suffered by London hospitals were relayed in detail. Bart's had in the last year cured 796 soldiers and civilians, and buried 116 more 'after much charge in their sickness'. St Thomas's had cured 825 and buried 116, while Bridewell, the hospital-cum-workhouse, had taken in 793 'Cavaliers and wandering soldiers, and other vagrant people', which had 'been very chargeable to the said Hospital, for apparel, sick diet, and surgery, besides their ordinary diet, and other provisions and charges expended about them, which could not be avoided by reason of their necessities'.[33] The importuning tone of this document lets slip that Cromwell's men were not quite as well provided for as he wished to make out, and the real state of affairs is revealed by Parliament's repeated acts to shake down every source they could think of for more funds. They exhorted church congregations to make 'a liberal contribution to this so pious and charitable a work', and when that failed they passed laws allowing their forces to seize bedding from private homes to put to use in hospitals (the law specified that it was to be returned and not 'imbezilled', but it is difficult to imagine how the goods were kept track of, or who wanted them back after they'd been used for hundreds of wounded soldiers).[34]

Outside London, the Royalists provided less comprehensive, less well-recorded care. Often forced into the provinces, they used field hospitals and local institutions; Parliamentarian newspapers mockingly described how the king ensured every town and village hospital in Crown-held areas were 'well stocked with maimed wretches'.[35] Eventually, injured and elderly troops loyal to the king would be better provided for, including in the grand setting of the Chelsea Pensioners' Hospital, but for the time being they had bitter complaints. In 1649, the year in which the king would lose his head, a disillusioned soldier named J. Rosworme published a pamphlet remonstrating with the Royalist forces for abandoning him to penury:

> What ingratitude is this, let the world judge; yea, judge your selves, ye worst of men; did I hazard life, limbs; and all that was dear to me, and do the richest of you grudge me a few shilling by the year, to buy me and mine food: is this your equity?[36]

Rosworme protested that he was angry with himself for ever having hazarded his life for such people, and for having prevented his men from plundering houses, when if he'd abandoned his scruples he might have made some money. Now, he was forced to do worse and turn highwayman in order to maintain himself and his family. Yet, he threatened darkly, 'I hope I shall find out a way to make you pay me against your wills, nor shall your dishonesty for ever help you.'[37]

* * *

The catastrophe of civil war proved once and for all the necessity for a functioning system of hospitals and welfare. However, it did not bring those things into being. In 1653, hundreds of sick and wounded sailors arrived in Portsmouth as casualties of the Anglo-Dutch wars, a conflict that would continue by fits and starts for over a century. The government knew by now that they needed hospitals to accommodate such men, and had ordered all the military hospitals to set beds aside, but sailors arrived far faster than places could be found for them, and Portsmouth and its neighbouring towns were inundated with maimed and diseased men: some dying on the streets, others crowding into port-side taverns where smallpox became rampant.[38] For years after Charles I's execution in 1649, Parliament tried in vain to foist the responsibility of caring for injured ex-servicemen, widows and orphans onto parish poor relief; when the exiled Charles II returned to claim the throne in 1660, he was no more successful.

By this time, military hospitals were functioning in a fairly orderly fashion. Places such as Ely House and the Savoy in London provided their occupants with clean linen and attentive nursing even if their cures were not very effective, and the records of daily fare from a military hospital in Edinburgh show soldiers receiving what was at that time deemed a nutritious diet: gruel, milk, beer, cheese, butter, bread and meat.[39] Things were less shipshape but still passable at the larger poor-hospitals.

Once again, John Ward's inimitable curiosity can help give us a layman's view. On a trip to London in 1665, he visited Bart's, which he referred to as 'the lame hospital, where are maintained near 300 men and women all infirm', and 'saw some of their wards very neat'. He also stopped in at Christ Church, the hospital-orphanage, and reported that 'the steward told me they maintained seven hundred in that hospital there and in the country and had not above 4000 [pounds] a year'.[40] He made a note to go to St Thomas's when he had more time. Ward's brief observations show that gentlemen with loose connections to the medical world were able to drop into hospitals at will and look at the patients there. It was an intrusion on the sick, but it demonstrates that, while madhouses and pesthouses were little more than detention centres, military hospitals were being viewed as forward-thinking medical institutions.

As for Bedlam, its continuing story writ large many of the (few) opportunities and (multitudinous) challenges faced by all sorts of hospitals at this time. The changes rushed through after Crooke's ruinous tenure did not have the transformative effect they had promised. Records of new rules brought in over the coming years show that, try as they might, the governors could never quite manage the hospital's staff, never mind patients. In July 1646, it was ordered that 'no officer or servant shall give any blows or ill-language to any of the mad folks'. In 1655, a call went out for governors who lived nearby to mount surprise inspections as often as possible 'to see how the lunatics are used', and two years later they decreed that men and women ought to be kept separate (easier said than done in the cramped old buildings).[41] But the renewed attention to hospitals during and after the civil wars had some benefits. In 1644, the ramshackle old building was extended to house both more patients and 'a house of easement' (a rather grand term for a latrine which was probably just a barrel or a hole in the ground).[42] As the city built up around the plot, the governors paid for five additional windows to try and regain some light, and numerous records show the committee attempting to find adequate clothes and bedding for their pauper residents.[43] However, it was Bedlam's next phase that would mark the birth of a modern institution, and signal the medical profession's intent – if not ability – to carve out a new era in hospital medicine.

* * *

When Bedlam was first designated as a hospice for the mad, it was designed to care for a few dozen patients at most. But the population kept growing, the need for care only increased, and by the mid-1660s, it was all too obvious that 'the hospital house of Bethlem is very old, weak & ruinous and too small and streight [confined] for keeping the greater number of lunatics therein'.[44] The upshot was that Bedlam badly needed a new home, and approval was given for a splendid new hospital at Moorfields, an area of London just outside the city walls where refugees from the Great Fire had camped out while their homes smouldered. Designed by the famed scientist Robert Hooke, it was finished in just two years, and in 1676, a poem was published in praise of the 'New Bedlam', dedicated to the master, governors and other 'noble benefactors of that splendid and most useful hospital'. In it, the author praised Bedlam for the purity of its air, its comfortable chambers, and the exemplary medical treatment offered to inhabitants.[45] It was, they mused, no wonder that some of the hospital's patients imagined themselves to be princes, when they lived in a building which so resembled a palace. Though this was an overstatement – after all, not many palaces had iron bars between one area and the next – the new building was certainly a vast improvement on the old one.

At the centre was a grand two-storey structure with a portico held aloft by vast stone pillars. This section contained the entrance hall, the apothecary room, staff accommodation and a visiting room for patients to meet their families. To each side were long galleries containing cells for patients, twelve by eight feet and each with natural light, a luxury in the old hospital. Men and women were separated at last, which prevented at least some of the sexual violence that had run rampant in Old Bedlam. What was more, the patients had access to outside space in sex-segregated exercise yards (though not in the large front garden, which was kept diligently free of lunatics). To underline the grandiosity of the project, the whole edifice was decked with a hotch-potch of ornamental stonework: ledges, carvings, even stone pineapples adorning the front wall.

The impact of New Bedlam on the lives of the people within was debatable. On one hand, the splendid new building attracted more visitors than ever before, arriving by the carriage-load to walk in the front gardens and roam the long corridors looking in on the inmates, like rare species in a particularly rowdy and disturbing zoo. This could be a welcome distraction for some patients, but it was more often an opportunity for young men to impress their friends by goading the most violent residents into displays of temper. Thomas Tryon, the writer on dreams and visions, importuned Bedlam's governors to stop the inferior officer admitting 'swarms' of nosy visitors:

> This staring Rabble seldom fail of asking more than a hundred impertinent questions. — As, what are you here for? How long have you been here, &c. which most times enrages the distracted person, though calm and quiet before, and then the poor creature falls a raving...whilst the wicked people, who think it a rare diversion, instead of trembling...fall a laughing and hooting, and so the poor distracted creatures become twice more fierce and violent than ever.[46]

On the other hand, the new hospital, under the supervision of physician Thomas Allen, made some strides towards affording its residents greater dignity. James Carkesse, who was confined to Bedlam for excessive religious 'enthusiasm' (he harassed and attacked members of other Christian sects), bitterly complained about his treatment at the hands of mad-doctors, but during his confinement he recorded how his friends had visited and given him a chair and some venison. That he was able to keep hold of these items implies Allen had his staff under better control than his predecessors. It was also to Allen's credit that when Richard Lower and other members of the Royal Society high-handedly demanded to be given a Bedlamite on whom to conduct blood transfusion experiments, he steadfastly refused (we will see what happened to the subject they eventually recruited shortly). After Allen's tenure, the hospital was reigned over by four successive generations of the Monro family. At their worst, their 'scientific' methods turned the institution from a dysfunctional hospital into a functional torture camp.

For all this, Bedlam's most important legacy was to be found outside its walls. The public interest in Bedlam – fuelled by its scandals more than its successes – meant that it was the first London hospital in 160 years to get its own purpose-built new premises. Once New Bedlam was up, however, the floodgates opened for hospitals to be founded, relocated or built from scratch. In 1682, work started on the Royal Hospital Chelsea, designed by Christopher Wren in the same grand, classical style. Greenwich Hospital followed a decade later, even finer and split in two so as not to spoil the view from Queen Mary II's house.* In 1733, St George's Hospital opened in Hyde Park, in a repurposed country house. St Thomas's Hospital in Southwark was almost entirely rebuilt between 1693 and 1720. Guy's Hospital for 'incurables' was founded in 1720, Westminster Hospital in 1719, the Middlesex Hospital in Marylebone in 1745 and the foundation stone for the London Infirmary on Whitechapel Road was laid in 1752. Outside London, hospital provision was still scanty, but new institutions sprang up in Northampton, Bristol, Edinburgh and York.

New Bedlam's era did not last. Pressure to complete the huge structure in record time forced its builders to cut corners, and by the 1800s it was crumbling, its inmates once again in need of new accommodation. The Moorfields building became the Imperial War Museum, prompting many platitudes about the madness of war. After several moves, the hospital is now based in Croydon, one of a dwindling number of dedicated psychiatric hospitals in a creaking system of mental health provision.

The turbulence and economic precarity of the sixteenth and seventeenth centuries acted like a social centrifuge, separating

* Both Greenwich and Chelsea were 'hospitals' in the old sense of the word, providing long-term accommodation for elderly or infirm ex-servicemen.

commingled communities and throwing the poor, disabled and infirm even further to the margins of society. Prejudice towards such people began long before the Renaissance and would continue long afterwards – texts from the 'enlightened' eighteenth century show the well-to-do complaining more vociferously than ever before about 'those creatures that go about the streets to shew their maimed limbs, nauseous sores, stump hands or feet, or any other deformity'.[47] However, the crisp conceptual separation between 'idle' and 'impotent', 'undeserving' and 'deserving' poor was a product of the Elizabethan era which came to shape welfare laws well into the twentieth century.

In this anxious climate, hospitals were left lamely trying to keep up. They expanded, sometimes rapidly, but need always grew faster. Yet the new policies they managed to implement – civilian governing boards, trained medical staff, extensive fundraising, expanded facilities – all helped transform them into hubs of medical research. As we turn to our final cohort of practitioners, however, we shall see that thirst for knowledge could be a dangerous thing.

Image of a man having liquid transfused into his veins, from *Clysmatica nova* (1667) by Johann Sigismund Elsholtz. The Wellcome Collection.

CHAPTER NINE

Remaking the World: Experimenters

It is 30 May 1667, London. The streets from Tower Hill to Ludgate ring with the sound of building and reek of smoke and sewage. They are cluttered with timbers, wagons and workers; everything has been upended. Houses lack occupants, gutted by the plague, still bearing the dreaded cross on their door. People lack houses, having fled the Great Fire with only what they could carry. Yet they have no choice but resilience, to hammer and saw and claw their way back to normality.

Meanwhile, in Arundel House just off the Strand, the world is being calmly dismantled. The scientists of the Royal Society are in the business of taking the universe apart to see its workings, and today they have a special visitor. Margaret Cavendish, Duchess of Newcastle, enters the room like a ship in full mast; rigged in fine silks, flanked by a flotilla of liveried servants.

The assembled scientists bustle to show Cavendish their experiments in chemistry and physics: 'First, that of weighing the air, which was done with a glass receiver of the capacity of nine gallons and three pints…Next were made several experiments of mixing colours. Then two cold liquors by mixture made hot. Then the experiment of making water bubble up in the rarefying engine, by drawing out the air.'[1] The showcase reflects their current obsession with understanding vacuums and pressure, but it is also carefully curated to omit less savoury objects of curiosity: the experiments on animals. And the ones on humans.

Allowing a woman into this enclave has generated controversy, and with it spectators, including Samuel Pepys. He doesn't much like the duchess, but he likes gossip and clothes, and she has clothes worth gossiping about; 'all the town-talk is now-a-days of her extravagancies, with her velvetcap, her hair about her ears; many black patches, because of pimples about her mouth; naked-necked, without anything about it, and a black just-au-corps [men's coat]'.[2] Now, he notes, her deportment is 'ordinary' and her clothes 'antick [chaotic]'. She has her own ideas about science and its uses, but for once she is content to put them aside. She only murmurs, 'she was full of admiration, all admiration.'[3]

* * *

Margaret Cavendish was a woman who split opinions. To her detractors – and there were many – she was 'mad Madge', a duchess with more money than sense, a freakish embarrassment to the aristocracy who insisted on dressing outlandishly, sticking her nose into places it ought not to be (the Royal Society) and airing her views on things of which she knew nothing (politics and science). Pepys called her 'a mad, conceited, ridiculous woman', adding that her husband William was 'an ass to suffer her to write what she writes to him, and of him'.* Her fellow gentlewoman Dorothy Osborne read her poems and concluded

---

* Despite this, Pepys went to see several of her plays, and when he first saw her in London, described her as 'very comely' (26 April 1667). In subsequent weeks he tried to get a closer look, but her finery and wild reputation attracted such crowds that he could see nothing.

'there are many soberer people in Bedlam'.[4] Yet to her supporters, she was a visionary polymath: playwright, novelist, scientist, philosopher and provocateur. She was a patron to many of the era's foremost thinkers, and the great chemist Robert Boyle was among many who fought for the Royal Society to admit her to a meeting. In her own right, she was a proponent of atomism (the belief that all things were made up of tiny indivisible particles), whilst also wary of the new science of microscopy, for how, she asked, could merely *looking* at things reveal their true nature? At times, her imagination took flight – her book *The Blazing World* was arguably the first work of science fiction, while *The Convent of Pleasure* imagined an all-female commune – but at others she championed common-sense science, delivered to the public in terms they could understand.[5]

Though Cavendish was an extreme example, the sixteenth and seventeenth centuries were full of characters whose interests roamed far and wide and made them experts in several subjects: housewife/medic; vicar/physician; surgeon/inventor; astrologer/apothecary. Moreover, the spirit of curiosity was somehow baked into this age. When so much of the world remained mysterious, experimentation was not the preserve of scholars in ivory towers, it was a part of life. After all, what else was the English Revolution but a great experiment, testing the thesis that a nation could exist without a monarch?

Medical practitioners were more invested than anyone else in experimentation, for good and ill. Each encounter with a patient was a roll of the dice in trusting that you had correctly identified their malady, then that you had chosen the right course of treatment, and finally that their body would cooperate. Medics of all kinds cheerfully described their practice as experimental

– John Ward wrote in his diary on many occasions that he was 'experimenting', sought out bear's testicles, ox's gall and a dead puppy to add to the cause, and added breezily that when physicians said they did a treatment 'tentatively', what they meant was 'experiment on human bodies'.[6] Still, there were experiments and experiments, and some made even hardened operators blanch. This chapter is about that latter kind. In particular, it is about the birth of a new scientific behemoth – the Royal Society – and their quest to discover what it meant to be human.

***

In 1503, when the Italian Renaissance was in full swing and the founders of the Royal Society not yet a twinkle in their great-grandparents' eyes, the Italian poet Elisio Calenzio wrote to his friend Orpiano:

> Orpianus, if you wish to have your nose restored, come here. Really it is the most extraordinary thing in the world. Branca of Sicily, a man of wonderful talent, has found out how to give a person a new nose, which he either builds from the arm or borrows from a slave. When I saw this, I decided to write to you, thinking that no information could be more valuable. Now if you come, I would have you know that you shall return home with as much nose as you please. Fly.[7]

The operation Calenzio described (and rather misrepresented) in this letter was an innovation in continental Europe, but not altogether new. It had first appeared in the ancient surgical writings of the Indian scholar Sushruta, thought to date from the sixth century BCE. Slowly, it traversed the four thousand miles from India to Greece and wound up in the hands of two Sicilian surgeons: Branca the elder and his son Antonio.[8]

It did so at precisely the right time. In Italy, bloody wars were being fought over territory, and a mysterious disease was threatening to wreak as much devastation as had the Black Death 150 years before. Syphilis – 'the pox' – appeared quite suddenly in 1494 and was likely spread through Italy by invading French armies. Where it originated remains a hot topic of debate among scholars, with some believing it to have been brought back from expeditions to the New World and others arguing it had been lying dormant in Europe. Nonetheless, the effect was the same – an epidemic of disfigurement caused by collapsed 'saddle' noses. Little wonder Calenzio was so excited at the prospect of a remedy, and so, it seems, was everybody else. Historians Martha Teach Gnudi and Charles Webster have shown that the Brancas taught their technique to German surgeon Heinrich von Pfolspeundt, and soon the nose graft was being discussed by various continental surgeons. Even the great Ambroise Paré weighed in on the debate, offering his opinion that the operation cost more pain and trouble than it was worth.[9]

Into this fray stepped a Bolognese surgeon named Gaspare Tagliacozzi. By the time Tagliacozzi learned of the nose reconstruction surgery, the procedure had suffered its own unfortunate transformation as it passed from one source to the next, and nobody could agree on the specifics of the operation (the rumour about getting a nose from someone else's flesh was false but incredibly persistent). Tagliacozzi thought anew, and set out a definitive method.

To begin the procedure, he explained, you will need a sharp knife, a lot of bandages and a very committed patient. This patient must first endure the skin of their upper arm being cut along the two long sides of a rectangle, and the knife slid under the skin to separate it from the underlying muscle, as if filleting a fish (you will require an assistant to hold the piece of skin taut with forceps, and probably another to hold the patient). It is the most difficult part of the operation, but the most essential, for it creates a flap still attached to the arm and its blood supply. Place a cloth soaked in egg whites underneath and bandage it for two weeks until it has somewhat healed.

When you revisit the patient, replace the egg-white dressing by stitching its long edge to that of a new piece, and pulling the whole thing through the wound. Leave it to heal a little more, and then cut along the third edge of the flap, considering carefully the size of the piece needed, which end to cut (nearest elbow or nearest shoulder), and whether to make a straight or curved incision. Patients will be eager for their new nose, but this phase cannot be rushed; lay the flap back down against the cloth, and leave it two weeks until the skin dries out and begins to curl upward.[10]

Now the reconstruction begins in earnest. The patient is to bring their arm up to their face, and the surgeon puts the flap from the bicep over the raw surface of the nose-stump, suturing the edges. Using a complicated system of slings and bandages, attach the arm to the face.

The patient must now remain in this position for twenty days, moving as little as possible. Ideally they should avoid talking, walking, laughing, sneezing. Shaving the head is a good idea – it will prevent lice, and therefore the urge to scratch. Naturally, staying like this will cause terrible pain in the arm and shoulder; if your patient complains too much, remind them of the stoicism of the ancient Greeks and give them some laudanum.

At the end of this period, one may – carefully – remove the bandages. Hopefully, what will be revealed is a graft between the skin of the arm and that of the face. Cut the skin flap along its fourth side, releasing it from the arm, and remove the sutures. Treat the wounds with zinc oxide, and by the fourteenth day, you should be able to 'forcefully' shake the graft without fear.[11]

With a rectangular flap of skin now attached to the face, the creation of a nose should be easy, though it takes several more weeks. Shape it to the desired proportions – bigger is better, for it will likely shrink in the first few years – and be sure to create nostrils (you will need to insert two small tubes while the skin is healing). If you are lucky, the colour of the arm skin will match that of the face, though it will likely be softer and paler. If not, you may find the new nose grows hair 'so luxuriant it must be shaved', but it is a small price to pay for a reconstructed face.[12]

Tagliacozzi's new operative method was published in 1597, two years before the author's death, in *De curtorum chirurgia per institutionem*. He might have expected that the audacious procedure would make his name. After all, Tagliacozzi himself boasted that

> the art makes it possible to restore noses so satisfactorily that sometimes nature is surpassed, and if you except the colour of the skin (which during the early days is not so healthy, yet there are means by which we can improve it), you would not in any way detect the made nose, as anyone can testify by the evidence of his senses who has

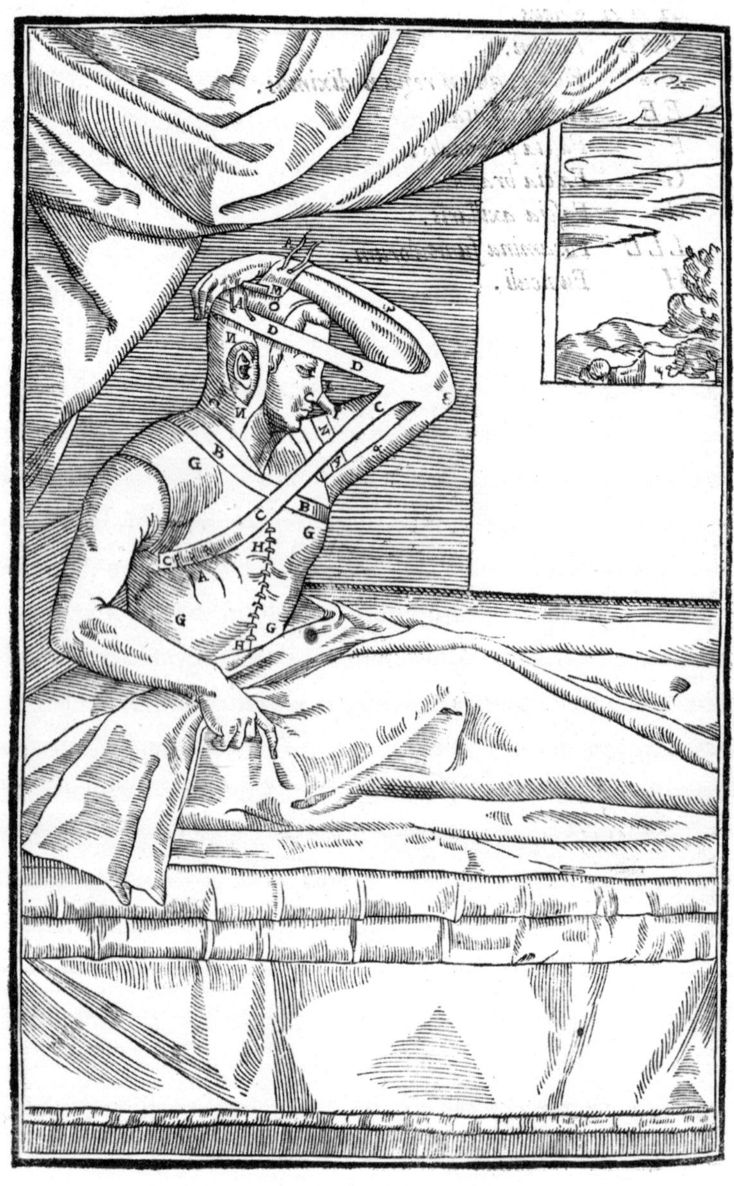

A man with his arm bound to his nose, from Gaspare Tagliacozzi's *De curtorum chirurgia.* The Wellcome Collection.

> seen the noses restored by me this year. For there have been several gentlemen...for all of whom we restored noses so resembling nature's pattern, so perfect in every respect that it was their considered opinion that they like these better than the original ones which they had received from nature.[13]

In a world where disfigurement was both common and socially ruinous, the notion of making faces anew was intoxicating. However, changing the face of medicine was not so easy. For a while, Tagliacozzi's method was much talked about, but as Paré had predicted, few patients or surgeons were brave enough to attempt such a difficult and dangerous cure. In the no-man's-land around this fascinating but increasingly stigmatised procedure, rumours began to swirl; but to understand how those rumours caught fire, we must return to England, where there was a group of daring experimenters making their own discoveries.

* * *

While physicians slowly received, digested and critiqued the news of Tagliacozzi's peculiar new surgery, in London, rather different kinds of experiments were taking place. These trials were conducted on live animals and corpses rather than patients; the men who thought them up saw themselves as scholars first, physicians second. Many key players were those with whom John Ward had shared dinners or trailed around after in Oxford. There was the wanly handsome Irishman Robert Boyle, fascinated by physiology and chemistry, feverishly dreaming up new schemes from his lodgings on Oxford High Street.* His friend Richard Lower was less intellectually gifted, but more practically minded, specialising in physiology and, later, physic. Thomas Willis started out selling medicines at the local market, though he'd come to be remembered as a gifted anatomist and physician, often collaborating with Lower. Young Robert Hooke, later famed for his work on microscopy, began his career as an assistant to first Willis, then Boyle, and worked on his own idea for a spring-regulated watch on the side. Christopher Wren was interested in absolutely everything; with John Wilkins, he built an eighty-foot telescope for observing the moon. Lawrence Rooke was professor of geometry, then astronomy – a decade later the group would be all together at his funeral, after he caught a deadly chill whilst stargazing.

They were a clever and ambitious bunch, sometimes quarrelsome (Hooke would later fall out with Isaac Newton spectacularly), but more often collaborative. Fifty years earlier, their sharp wits might have been best used earning themselves preferment at court, or perhaps an influential job in the civil service. Now, however, society and the economy were changing, and the nation needed innovators, not aristocrats. Oxford was the perfect incubator, a city of lively coffee-houses, well-stocked cellars and skilled instrument-makers accustomed to the strange requests of the dons. Wilkins, the warden of Wadham College, had been sent to Oxford by his brother-in-law Oliver Cromwell

---

* John Ward wrote admiringly that Boyle sustained his intellect with a strict physical regimen: 'Mr. Boyle never drinks any strong drink or wine: he every morning eats bread and butter with powder of Eyebright spread on the butter: his supper is water gruel and a couple of eggs; his dinner is of mutton or veal, or a pullet…he reads a chapter in Greek [and] one in Hebrew every morning.' (Ward v13, f.27r)

to help temper the Royalist sympathies gathering there, but he soon found he had a gift for bringing like-minded people together rather than forcing them apart. Under his influence, the Oxford Philosophical Society was formed in the early 1650s, a 'clubb' of about thirty members, whose first business was to find out what experiments were being done and where.[14] In Wadham's newly planted formal gardens, Wilkins set up scientific equipment, as well as curiosities such as a 'talking' statue and a transparent beehive. The young men bragged, joked, argued and compared notes into the night.

In 1658–9, the club relocated to London. Their commitment to discovering the laws of nature had provoked hostility from Oxford and Cambridge, whose curricula were firmly based on Aristotelian principles. It proved a savvy move, despite their difficulties in securing a place to meet (they'd expected to be permitted to use the College of Physicians' premises. The physicians thought differently). Here too they saw the opportunity to turn their group from a loose collection of like-minded individuals into a society, with official membership, meetings and its own scientific journal. In 1660, Charles II returned from exile to take up the throne, and society members immediately approached the monarch with assurances that a royal charter would enhance the country's reputation without requiring funds from the impoverished ruler. Two years of administrative wrangling later, they had the seal of approval.

Membership of the society now began to swell, from 11 in 1660 to over 120 in 1663, each carefully chosen by the founders for both brains and connections. They included the aristocrats Sir Kenelm Digby (advocate of the cure-all 'weapon salve', who would later make outlandish claims about the Tagliacotian nose operation) and George Villiers (son of James I's notorious 'favourite' and a close ally of Charles II), as well as a scattering of European intellectuals such as the Dutch mathematician-physicist Christiaan Huygens and French alchemist Nicolas le Febure. They experimented with anything and everything: microscopes, telescopes, the generation of insects, taking the temperature of deep mines and mountaintops, freezing the blood, the refraction of light by different substances, echoes, musical harmonies, teaching deaf people to speak, the speed of a bullet, the motion of pendulums, the depth of the sea, the strength of winds, the density of saltwater, the heat of fires, the use of the spleen in animals, the anatomy of the brain and of chameleons, dogs, trees and fishes, the melting point of metals and the possibility of 'inchanting' a spider by drawing a circle of unicorn horn around it.[15*]

Admittedly, it was not round-the-clock genius. The meticulously kept minutes of their meetings show some days when these highly educated men earnestly discussed the possibility of making a cart with legs instead of wheels, or spectacles for seeing in the dark. Others were just dull, as these notes from a gloomy January gathering show:

> Mr Boyle mentioned, that he had been informed, that the much drinking of coffee did breed the palsy [weakness].
>
> The Bishop of Exeter [Seth Ward] seconded him, and said that himself had found it dispose [him] to paralytical effects…

---

* Unicorn horn in this period described what we would now recognise as narwhal horn or elephant ivory. Supposedly the spider would not cross the circle made by the unicorn horn, but in this event it repeatedly scuttled straight through it.

> Mr Graunt affirmed, that he knew two gentlemen, great drinkers of coffee, very paralytical.
>
> Dr Whistler suggested that it might be enquired, whether the same persons did take much tobacco.[16]

At other times, members' interests led them to darker endeavours, dispassionately performing horribly cruel experiments on animals. They poisoned or vivisected scores of cats, dogs and birds to test the action of drugs or see which organs one could live without. A particular fascination was putting various creatures into a dome from which the air was mechanically removed, and noting how long it took them to die (a 'lucky' few were resuscitated, but probably only to be reused in another trial).

It was as dangerous as it was exciting, for Boyle and his colleagues viewed themselves as pioneers at the vanguard of a glorious new age for science. Provided that they followed the guiding light of rationality, they could do no wrong, for was not the world created by God? Were its mysteries not puzzles set by the Almighty for men to unravel? Belief in their purpose made them bold, but they were about to meet the limits of their power.

The transfusionist: Richard Lower

Among the Oxford set who went on to membership of the Royal Society, Richard Lower was neither the most famous, nor the cleverest. He did not have the restless genius of Wren, the imagination of Boyle or even the cantankerous doggedness of Hooke, so he has been largely forgotten in the history of science. What he did have, however, was a strong stomach, deft hands and a shrewd eye. In February 1665, aged thirty-four, Lower was still in Oxford whilst many of his some-time collaborators had moved to London. As assistant to Thomas Willis, he'd become known as a gifted physiologist, and Willis sang Lower's praises in his 1664 *Cerebri Anatome*, praising 'the sharpness of his scalpel and his intellect'.[17] Now, however, Lower was ready to make his own mark on the scientific community, and at a meeting of like-minded Oxonians, he debuted an exciting new experiment: the transfusion of blood from one dog into another.

The procedure had taken Lower months to perfect. In a letter to his friend Robert Boyle, he described how he'd initially tried to get the blood to flow from one animal's jugular into the other, but it kept clotting. The problem was compounded by the fact Lower had to use an elaborate system of quills to transfer the blood. One such quill, though not Lower's own, survives in the International Museum of Surgical Science in Chicago. The fragile joins over just a few inches of quill-tube show how difficult it must have been to extend the system over a longer distance. Eventually, he realised that a transfusion from carotid artery to jugular vein would be more successful, but it still involved carefully isolating the vessels, ligating them at the appropriate points and stoppering the quills until everything was arranged. It also involved procuring a number of ill-fated dogs, for in these initial experiments, Lower wished only to get one animal's blood into another, a task which he regarded as complete when the donor dog died. The survivor dog, he reported, 'runs away lively and strong' and, if it had any sense, looked for the first opportunity to escape Lower's clutches.[18]

News of Lower's experiment soon spread; John Ward dutifully recorded it in his diary, and Boyle encouraged his friend to show his work to the Royal Society.[19] The next year, their journal *Philosophical Transactions* gleefully reported that 'this experiment, hitherto looked upon to be of an almost unsurmountable difficulty', had now been demonstrated to great acclaim in both Oxford and London.[20] Nonetheless, whilst the trial was impressive, it wasn't clear how it was *useful*. In a blink-and-you'll-miss-it passage in his *Tractatus*, Lower suggested that infusing animal blood into people who'd suffered haemorrhage might save lives (neither Lower nor anybody else had any notion of blood compatibility).[21] Strangely, however, this potentially revolutionary insight was overlooked. Instead, Boyle urged Lower to focus on the *nature* of blood, its intrinsic qualities and mysterious powers. In a public letter, he proposed a number of further experiments. Could transfusing the blood of a cowardly dog into a fierce one make the recipient tamer? Would the blood of a fighting dog put into a bloodhound make the latter's sense of smell weaker? Might a dog be kept alive without eating, if it was frequently injected with the chyle (liquid digested food) from another? Could a diseased hound be cured by exchanging its blood with a healthy animal, or vice versa? Would a small dog grow bigger, infused with blood from a larger counterpart? What about transfusions into pregnant bitches – would their puppies have the tinge of a different breed? Lastly, Boyle enquired 'Whether the operation may be successfully practised, in case the injected blood be that of an animal of another species, as of a calf into a dog, &c. and of a cold animal, as of a fish, or frog, or tortoise, into the vessels of a hot animal [mammal], and vice versa?'[22]

By now it was obvious which way the wind was blowing. The quest to understand how blood worked would inevitably end with human transfusion. However, Lower was beaten to the punch. In

1667, he learned to his chagrin that a Frenchman called Jean-Baptiste Denys had managed to infuse blood from a calf into a sixteen-year-old boy and a middle-aged man, with apparent success. Adding insult to injury, Denys's account of his triumph completely failed to acknowledge Lower's work in this area (or that of fellow physicians Edmund King and Thomas Coxe), and instead claimed he'd got the idea from an obscure Benedictine monk.[23] Lower scrambled to catch up. He was by now resident in London, gradually establishing himself as the city's most eminent physician, and determined to use every lever in his power to reclaim his place as Europe's most advanced transfusionist.

Finding a person willing to have animal blood put into their body was more difficult than Lower and his co-experimenter Edmund King had imagined. They approached the keeper of Bedlam, Thomas Allen, to get themselves a madman. The rumours were that Denys's next trial would be on someone with madness, so they could get ahead of their rival, and in addition, the residents of Bedlam tended to be poor, friendless individuals, unlikely to cause problems if things went wrong. To his credit, however, Allen refused their request. The residents of Bedlam suffered many indignities, but he was determined that being experimented on should not be one of them. Lower and King were forced to cast about for an alternative, and they found one in the form of Arthur Coga, a thirty-two-year-old divinity graduate and member of the same church as their old friend John Wilkins. Coga was said to be a little mad, but his bigger problem was poverty. Lower persuaded him to participate for the fee of just twenty shillings. The planned experiment caused a stir in the city, and once again Samuel Pepys was on the periphery, recording his impressions:

> the College [Royal Society] have hired [Coga] for 20*s.* to have some of the blood of a sheep let into his body; and it is to be done on Saturday next. They purpose to let in about twelve ounces; which, they compute, is what will be let in in a minute's time by a watch. They differ in the opinion they have of the effects of it; some think it may have a good effect upon him as a frantic man by cooling his blood, others that it will not have any effect at all. But the man is a healthy man, and by this means will be able to give an account what alteration, if any, he do find in himself, and so may be useful.[24]

On 23 November 1667, the experiment began at the Royal Society's base, Arundel House. This time, Lower and King wisely did not attempt to transfer blood directly from the sheep into Coga. Instead, they bled the donor animal into a bowl, then let six ounces of blood from Coga. Using Lower's tried-and-tested quill system, they put the sheep's blood into Coga's vein. They waited nervously...nothing happened. Coga, sensing another payday on the horizon, claimed to feel much refreshed, proposing another transfusion within the next few days (seeing him a few days later, Pepys was less convinced, judging him still 'cracked a little in his head').[25] Cautiously, Lower and King repeated their trial in December, this time attempting to transfer a full fourteen ounces of blood. Once again, very little effect could be seen, though Coga gamely continued to play up the benefits of the procedure. Though the members of the Royal Society did not know it, putting such a foreign substance into the bloodstream of a human, even in relatively small quantities, should have been fatal.[26] The fact that Coga lived probably indicates that hardly any sheep's blood was actually entering his veins (some observers suspected as much at the time, and called for both Coga and the sheep to be weighed before and after the transfusion).

Lower was the man of the moment, granted the honour of being put in charge of experiments at the Royal Society. The future for blood transfusion also looked bright. While Lower and King had been experimenting on Coga, Denys had managed to repeat the procedure on a violent madman named Antoine Mauroy, forcibly infusing him with calf's blood after which he became 'of a very calm spirit, performs all his functions very well, and sleeps all night long without interruption'.[27] George Acton, a shadowy figure who described himself as the king's alchemist, wrote excitedly of the medical possibilities of transfusion. Knowing the powerful properties of animal ingredients in regular remedies, he believed it evident that infusions of the blood of a male goat would 'certainly' cure pleurisy and pneumonia; the blood of an ox, dysentery; of a cat, epilepsy and herpes; of a fox, kidney stones; and of a stag, gout.[28] Perhaps the blood of long-lived animals such as eagles might even extend the lifespan of man. A new world of medical possibilities was emerging.

Alas, the shining new dawn would soon turn dark. In spring 1668, Mauroy went back to his violent ways and Denys attempted another transfusion.* This time, Mauroy died, and before long his widow was loudly proclaiming that it was all Denys's fault. Meanwhile in London, Coga had turned on his paymasters. Perhaps influenced by enemies of the Royal Society, he made

---

* Sources vary on whether they actually managed to get the blood into Mauroy, and whether he died immediately or the next day.

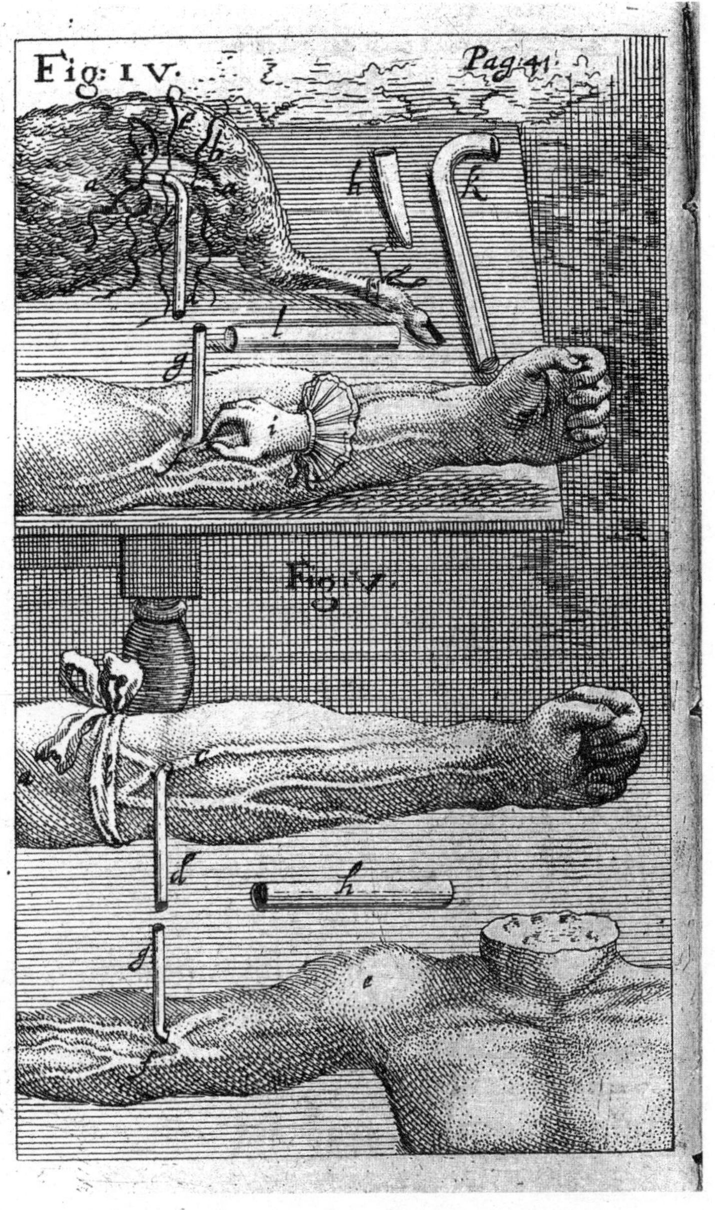

Diagram showing transfusion of sheep and human blood, from *Clysmatica nova* (1667) by Johann Sigismund Elsholtz. The Wellcome Collection.

alarming assertions, which were published in physician Henry Stubbe's anti-Royal Society text of 1670:

> Your creature (for he was his own man till your experiment transformed him into another *species*) amongst those many alterations he finds in his condition…finds a decay in his purse as well as his body…'Tis very miserable, that the want of natural heat should rob him of his artificial too; But such is his case; to repair his own ruins (yours, because made by you) he pawns his clothes, and dearly purchases your sheep's blood with the loss of his own wool. In this sheep-wrecked vessel of his, like that of *Argos*, he addresses himself to you for the Golden Fleece. For he thinks it requisite to your Honours…to transform him without as well as within. If you oblige him in this, he hath more blood still at your service, provided it may be his own, that it may be the nobler sacrifice.
>
> The meanest of your flock,
> Agnus Coga[29]

Coga's language was flowery, but his claim was ugly; he'd been made the sacrificial lamb to Lower and King's ambition (*Agnus* means 'lamb' in Latin), and when they'd finished exploiting him, they tossed him aside, leaving him so poor he had to sell his clothes. Furthermore, he hinted at something worse – that in infusing him with sheep's blood, the experimenters had made him a *creature*.

It was a topic ripe for satire, and Thomas Shadwell's play *The Virtuoso* soon stuck the boot in. The transfusionists Lower and Coxe were represented in the madcap figure of Sir Nicholas, a man whose other endeavours include learning to swim by watching and imitating a frog, whilst studiously avoiding any contact with the water. The blood transfusion experiment has, Sir Nicholas admits, killed 'four or five' subjects before it finally 'succeeds' by creating a monstrous human-animal hybrid:

> The patient from being maniacal, or raging mad, became wholly ovine or sheepish; he bleated perpetually, and chewed the cud: he had wool growing on him in great quantities, and a Northamptonshire sheep's tail did soon emerge or arise from his anus.[30]

Only a few years after it began, the blood transfusion project was in tatters. Nervous authorities banned any more such experiments in France, and the Royal Society wisely elected to cut this line of enquiry short. (It was not revived until the nineteenth century, and remained a risky business until human blood types were discovered by the Austrian doctor Karl Landsteiner in 1901.) Lower stopped attending Royal Society meetings, and eventually resigned his fellowship. Yet he was too canny and ambitious to be downtrodden for long. His anatomical knowledge made him an excellent doctor, and he was appointed the royal physician in 1675; the Oxford antiquary Anthony Wood described Lower as 'the most noted physician in Westminster and London…no man's name was more cried up at court than his'. His experimental subject was less fortunate. Melancholy, friendless and poverty-stricken, Arthur 'Agnus' Coga faded from the historical record, never to be seen again.

***

While Lower, King, Boyle and Coxe were flexing their intellectual muscles in transfusion experiments, news of Tagliacozzi's nose graft had traversed the continent and reached English shores. The Royal Society scientists would certainly have known about this strange operation. Like their own endeavours, it begged questions about what belonged to the body and what was interchangeable; indeed, they would later attempt their own transplant experiments, grafting various items including claws and human teeth onto the head of cockerels.[31] Yet, the transfusionists studiously avoided making the connection. They considered themselves genteel denizens of science, and in the hundred years since Tagliacozzi had made nose grafting famous, the operation had fallen into disrepute for being too difficult, too dangerous and simply too *weird* to countenance. There was also a graver problem, summed up in 1669 by one of the Royal Society's more eccentric members, Sir Kenelm Digby. In a book about his invention of 'sympathetic powder', he confidently asserted,

> I will say nothing of artificial noses, made of the flesh of other men, to remedy the deformity of those, who by an extreme excess of cold, have lost their own: which new noses petrify, as soon as those persons, out of whose substance they were taken, come to die…For, though this be constantly avouched by considerable authors, yet I desire you to think that I offer you nothing which is not verified by solid tradition; such, that it were a weakness to doubt of it.[32]

There is much to unpack here, starting with Digby's insistence that he believed nothing without 'solid tradition'. Though an energetic and enthusiastic natural philosopher (as well as courtier, naval officer and advocate of religious toleration), even Digby's best friends would have admitted he was credulous by nature. He believed fervently in the power of sympathy, in which atoms from a particular source retained a connection with that source, and in evidence he cited people who got ill from a dog bite if the dog later contracted rabies, the power of a toad to draw out pestilence when kept in a box around one's neck (toads were widely believed to be venomous), and the 'fact' that farmers repelled boys from defecating in their fields by applying a hot poker to the offending substance. Still, the rumour Digby was repeating *was* oddly persistent. Ever since the Brancas first started grafting noses, there had been mutterings that rich patients were paying poor people – or even forcing their servants – to have a chunk of their flesh cut off to supply the graft, rather than enduring the procedure themselves.

Tagliacozzi tried hard to smother this idea. As he pointed out, grafting between two people would be 'doubtful, if not entirely vain', since they would need to be tied together for weeks on end.[33] Despite his protestations, the rumour was not extinguished. Instead, it grew even more audacious, as other physicians started to claim that flesh grafted on from another body remained affiliated to its source, and would start to rot when the donor died. Jean Baptiste van Helmont (father of Lady Conway and Henry More's esteemed physician Francis van Helmont) claimed that a man from Brussels had gained a new nose from Tagliacozzi after losing the original in a duel. He feared having his own arm cut open, so hired a porter, from whose flesh 'at length he digged a new nose'. However:

> About thirteen months after his return to his own country, on a sudden the ingrafted nose grew cold, putrefied, and

> within a few days, dropped off. To those of his friends, that were curious in the exploration of the cause of this unexpected misfortune, it was discovered, that the Porter expired, near about the same punctilio of time, wherein the nose grew frigid and cadaverous.[34]

Material like this was too good for any self-respecting wit to ignore. Lady Hester Pulter penned a poem to (the famously noseless) Sir William Davenant, offering him a chunk from her leg to make a new nose, if he could only take care not to lose it to syphilis again.[35] Between 1663 and 1678, Samuel Butler released the satirical epic *Hudibras*, in which he reported confidently that

> So learned Talicotius from
> The brawny part of Porter's bum
> Cut supplemental noses, which
> Lasted as long as parent breech:
> But when the date of nock was out,
> Off drop'd the sympathetic snout.[36]

The well-reputed surgeon Alexander Read attempted to bring Tagliacozzi back into the medical fold in 1687 by including a translation of *De curtorum chirurgia* in his own *Chirurgorum comes*. But it was too late. The genie was out of the bottle, and nose transplants continued to be the butt of jokes well into the eighteenth century.

* * *

The parallel careers of these two experiments – both ending in ignominy – might seem to have little to do with the medical experiences of everyday people. After all, neither blood transfusion nor nose grafting was widely adopted by seventeenth-century medics (though both would be revived in the twentieth century – an almost exact copy of Tagliacozzi's nose operation was used to supply new noses for men disfigured in the First World War).

Yet these radical experiments, and the outraged, fascinated gossip they provoked, changed the medical landscape. Working out the mechanics of grafting on a skin flap or transfusing blood produced far more detailed understandings of the circulatory system, which would eventually contribute to the decline of therapeutic bloodletting. Only fifty years earlier, before William Harvey's *de Motu Cordis* revealed the workings of the heart and lungs, medics believed that blood moved between the venous and arterial system by passing through invisible pores in the ventricles. The role of the heart was merely to warm the blood, while that of the lungs was to cool it and impart a spiritual essence known as 'pneuma'. Now, the Italian Marcello Malpighi had shown the existence of capillaries, and Thomas Willis had traced the route of the carotid arteries into the brain. There were philosophical ramifications too. Seventeenth-century philosophers – René Descartes, John Locke and Henry More, to name a few – became obsessed with the question of what made an individual *themselves*. If a person had a replacement nose, did they remain the same person? What about new blood? What about a new *body*?

Scandal and ridicule were as important as glory and approbation. The Royal Society, with its lofty ideals and devoted membership, was an engine for scientific progress. But as its more outlandish experiments reveal, it also needed the court of public opinion to keep it in check. In service of their curiosity, Lower, Boyle and others glossed over the dubious parts of their work, in which they paid little heed to the welfare of their experimental subjects or the possible consequences of their findings. Without the watchful eye of the satirists, they might have become a group of Frankensteins, thoughtlessly creating monsters.

There was a final twist in the tale of this era's strange, brave experiments. After Antoine Mauroy died following a third attempted blood transfusion, the doctors, Denys included, wanted to open his body and puzzle out what had happened. Though Mauroy's widow, Perrine, had previously been happy for the doctors to try their experiments – in fact, Denys claimed she had begged them 'by unwearied clamour' to try the procedure again – she now changed her attitude. She violently opposed postmortem, and as Denys later told it, 'when we threatened her, that we would return next morning, and do the thing by force, she caused her husband to be buried an hour before day, to prevent our opening of him'.[37] Denys was suspicious, but knew he was on thin ice; soon enough, a court case was brought against him. However, he asserted that before the case was due to come to trial, Perrine approached him, confessed that Denys's enemies had given her money to testify against him, and offered to switch her allegiances for cash. The doctor refused and promptly assembled his own countersuit, finding more than he'd dared imagine. Several of the Mauroys' neighbours attested that in the weeks before Antoine Mauroy's death, they'd heard the couple arguing violently, and seen Perrine giving some liquid to a cat, which died. It was scanty evidence, not enough to have convicted someone of a higher rank, but just as in the Overbury poisoning case, it was far easier to condemn an impoverished person than look too closely at the wealthy individuals at the heart of the matter. The judge found Perrine guilty, and sent her to the grim Grand Châtelet prison on the banks of the Seine. Furthermore, he ordered that all transfusion experiments be suspended until the (distinctly conservative) medical faculty of the University of Paris had a chance to look into the matter. In effect, it was a ban. The faculty had always opposed transfusion, and they saw Denys's trials as typical of his personality: hasty, overambitious, insufficiently awed by his elders and betters.

If Perrine Mauroy had not been persuaded to lie about her husband's death, medicine for the next two hundred years might have looked very different. Incompatible blood was bound to kill a recipient sooner or later, but if the victim's family was less motivated to publicise the fact, the experiments might have evolved, and human-to-human transfusion might have saved lives. As in the case of the Chamberlen family and their secret forceps, personal agendas were as powerful a force in shaping the medical landscape as any grand institution or scientific breakthrough. Nonetheless, the seventeenth century was the age in which doctors and scientists turned to first principles in their experimentation – the beginning of what would later be called the 'Scientific Revolution'. Though their ideas were sometimes bizarre and their methods often unethical, they understood that to know how to cure the human body they needed to comprehend its workings; and the way to start was by questioning *everything*.

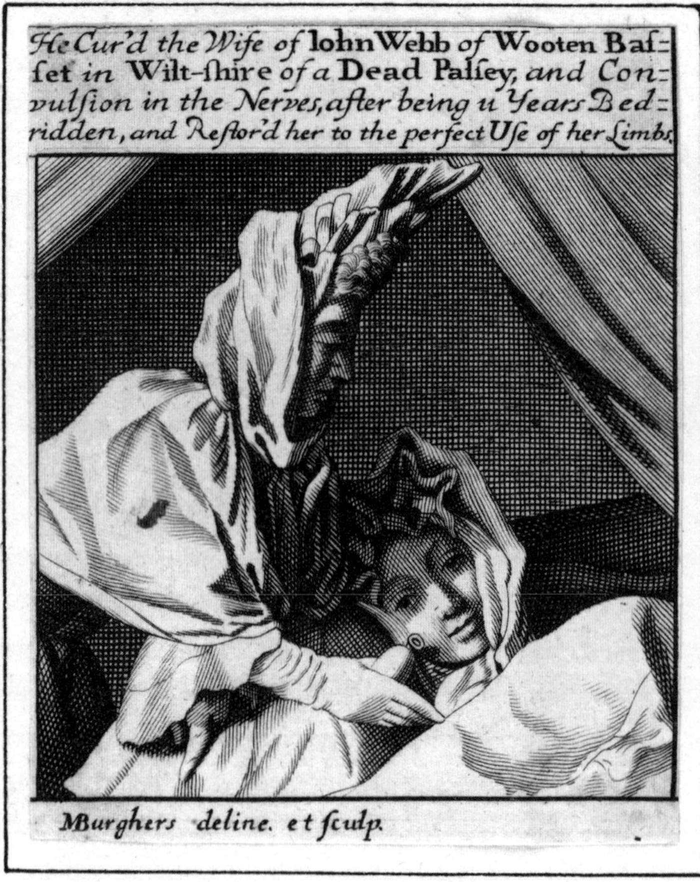

Mrs John Webb, being nursed when sick in bed with 'a dead palsey, and...convulsion in the nerves', before being cured by Sir William Read. Engraving by M. Burghers, c. 1700. The Wellcome Collection.

Conclusion

On 13 June 1704, Anne Chaytor of Croft, a picturesque village on the bank of Yorkshire's River Tees, wrote to her father William about her mother, Peregrina. The doctor had come to see Peregrina the previous night, but things had not gone well:

> The doctor and I had a sad bout with my mother last night about the diascordium and though she begged as if it had been for [her] life he positively forbid it and charged me against it so I was forced to sit up with her and intend to sit up again tonight for I dare not trust anybody, she is so earnest for it that they tell me they can't defeat her. 'Tis a hard part I have to act every way but I'm resolved to go through with it and run the hazard of being sick or anything rather than see her ruin herself. I've humoured her too long and now I've got the maisterhead [upper hand] I'm resolved to keep it. She is worse to wean from her sleepy doses than any child from the breast which is wonderful [i.e. astonishing] in a woman of sense. God grant I keep my health and I hope all will do well. She is more easy this afternoon and that is what I'm resolved if it be possible she shall keep to.[1]

Peregrina was sick, but not with pox or plague. Diascordium was a well-known medicine, once recommended by Nicholas Culpeper for hastening labour and reducing morning sickness. Its long list of ingredients included red roses, cinnamon, gentian, dittany, tormentil root, bole armoniac and, crucially, opium. What Peregrina was suffering was addiction to opiate painkillers, and Anne was witnessing the terrible effects of withdrawal.

Anne and Peregrina's story speaks to the uncanny way in which early modern patients and medical practitioners can seem both close at hand and unfathomably distant – at one moment, our next-door neighbours and the next, strangers in a strange land. Peregrina's affliction is one with obvious twenty-first-century correlates, but in her own time it was a mysterious disorder, and she had to find her way back to health aided only by Anne's determination and intuitive understanding of the need for abstinence. Anne's collaboration with the doctor also shows the nuance of this relationship. She saw herself as working with the physician rather than simply following his instructions, and he knew the cure could not succeed without her nursing skills.

This was the product of a unique point in history, when medical practice was growing in confidence and authority, but also becoming ever more diverse and varied. Seventeenth-century medical practitioners had access to substances and techniques they could only have dreamed of a hundred years before. They could make a man sweat until his teeth fell out, pull a baby from the womb, create prostheses that allowed the lame to walk and the disfigured to show their faces, and name every muscle in the human body. Yet there was still so far to go to understand why plagues spread and how to stop them, how to keep an injured soldier from bleeding to death in one's hands, what medicine to give an ailing infant or how to relieve pain during an amputation.

These push-and-pull forces kept the medical market growing in every direction. Patients and their families did not choose between quacks and doctors. They made their selection from physicians, domestic healers, apothecaries, surgeons, dentists, bone-setters, midwives, faith healers, prosthetists, mad-doctors, astrologers, scientists and mountebanks. Medicine became a true business in this period, with competition and advertising, governing bodies and trade secrets. It was not a level playing field; what kind of care you could seek (or what kind of practitioner you might become) was constrained by class, gender and, above all, wealth. Yet here too was the beginning of the state-sponsored care that would eventually become the welfare system and the National Health Service. Looking after the poor and sick became a secular duty as well as a religious one.

Medicine's transformation was sparked by the intellectual revolution of the Renaissance, driven by the printing press, the rise of continental universities, a loosening of the Church's grip on scholarship and the revival of classical learning. However, it also needed fuel. The hardships of life in the sixteenth and seventeenth centuries forced medical practitioners to innovate. Surgery would not have advanced at the pace it did without bloody conflicts supplying a stream of bodies in need of repair. Hospitals might never have proliferated if civil war had not made them urgently necessary. Domestic medics were driven to perfect their craft because so few other options were open to clever women. And all those who made their living from medicine fought hard for their customers, because the cost of living kept on rising.

Far removed as the blood-letting and bone-setting of old may seem, the legacy of this period remains with us. Important breakthroughs were made during these centuries: William Harvey discovered the mechanics of the circulatory system, Andreas Vesalius's magnificent *Fabrica* changed our understanding of anatomy forever, and new medicines made from chemicals and exotic New World substances entered apothecaries. However, the most important changes were systemic. Here was the birthplace of osteology, pharmacy, psychiatry, obstetrics and many other medical specialisms which still endure. For that matter, formal medical training can itself be traced back to this era. The Royal College of Physicians and Royal College of Surgeons (who separated from the Company of Barber-Surgeons in 1745) remain the United Kingdom's foremost bodies for regulating and educating medical professionals. What is more, the sixteenth and seventeenth centuries saw a sea-change in medical *ambition.* Practitioners of all kinds pushed the boundaries of their practice, urged on by competition and collaboration, filled with the energy of this extraordinary time.

But these were broad shifts, more visible now from the historian's vantage point. On the individual level were countless men and women much like ourselves, neither heroes nor villains but clever, fearful, selfish, kind, greedy, god-fearing, imaginative people doing their best to make a living and a name for themselves, looking to get a leg up wherever they could. And at every turn, those who mattered most but spoke least: the patients. As William wrote back to Anne, 'you now make use of the best medicine…resolution and courage'.

Acknowledgements

The idea for this book was conceived in 2021, and like good Renaissance gossips, many people have helped me through the long labour of bringing it into the world. Dr Amie Bolissian McRae was invaluable in tracking down lost references and obscure facts, a truly adept historian. Dr Emma Marshall was kind enough to send me the 'diascordium' letter with which the book concludes. For details about the asylum-keeper John Ashbourne, I am indebted to Andrew Mason, who now lives in the Old Rectory once occupied by Ashbourne and his lunatics. My colleagues in the Department of English Literature at the University of Reading have been supportive throughout. My agent at Watson Little, Donald Winchester, was invaluable in bringing the theme of the medical marketplace to the fore, and at every stage since. Thanks must also go to my editor at Oneworld Publications, Sam Carter, and his colleagues Rida Vaquas and, especially, Hannah Haseloff. Their editorial suggestions have made the work better, and me a better writer. My friends and fellow writers Emma Cullen and Leonie Fraser have also read many pages, and borne with my peculiar interests with good grace. The hero of the piece is my husband Sam, who now knows more about wandering wombs and amputated limbs than he ever could have wished for.

Further Reading

Prologue

Noga Arikha, *Passions and Tempers: A History of the Humours* (New York: HarperCollins, 2007).

Jonathan Sawday, *The Body Emblazoned: Dissection and the Human Body in Renaissance Culture* (Abingdon: Routledge, 2013).

David Hillman and Carla Mazzio, eds, *The Body in Parts: Fantasies of Corporeality in Early Modern Europe* (New York: Routledge, 1997).

Health Begins at Home: Domestic Healers

Elaine Hobby, 'A woman's best setting out is silence: the writings of Hannah Wolley' in Gerald MacLean, ed., *Culture and Society in the Stuart Restoration: Literature, Drama, History* (Cambridge: Cambridge University Press, 1995), pp. 179–200.

Elaine Leong, 'Making Medicines in the Early Modern Household', *Bulletin of the History of Medicine* (2008), 82.1, pp. 145–68.

Joanna Moody, *The Private Life of an Elizabethan Lady: The Diary of Lady Margaret Hoby, 1599–1605* (Stroud: Sutton Publishing, 1998).

Michelle DiMeo and Sara Pennell, eds, *Reading and Writing Recipe Books, 1550–1800* (Manchester: Manchester University Press, 2013).

Elizabeth Freke (ed. Raymond Anselment), *The Remembrances of Elizabeth Freke 1671–1714* (Cambridge: Cambridge University Press, 2001).

Cornering the Market: Physicians

Daniel Defoe (ed. Cynthia Wall), *A Journal of the Plague Year* (London: Penguin, 2003).

Harold J. Cook, *Trials of an Ordinary Doctor: Joannes Groenevelt in Seventeenth-Century London* (Baltimore: Johns Hopkins University Press, 1994).

Roy Porter, *Bodies Politic: Disease, Death and Doctors in Britain, 1650–1900* (New York: Reaktion Books, 2013).

Ian Mortimer, *The Dying and the Doctors: The Medical Revolution in Seventeenth-Century England* (Woodbridge: Boydell and Brewer, 2009).

Right Place, Right Time: Surgeons

Joanna Bourke, *The Story of Pain: From Prayer to Painkillers* (Oxford: Oxford University Press, 2014).

Alanna Skuse, *Surgery and Selfhood in Early Modern England: Altered Bodies and Contexts of Identity* (Cambridge: Cambridge University Press, 2021).

David Schneider, *The Invention of Surgery: a history of modern medicine from the Renaissance to the implant revolution* (London: Coronet, 2021).

Arnold van de Laar, *Under the Knife: A History of Surgery in 28 Remarkable Operations* (London: John Murray, 2018).

Medicine for the Masses: Apothecaries

Benjamin Woolley, *The Herbalist: Nicholas Culpeper and the Fight for Medical Freedom* (London: HarperCollins, 2012).

Louise Hill Curth, ed., *From Physick to Pharmacology: Five Hundred Years of British Drug Retailing* (Farnham: Ashgate, 2006).

Andrew Wear, *Knowledge and Practice in English Medicine, 1550–1680* (Cambridge: Cambridge University Press, 2000).

In the Beginning: Midwives

Adrian Wilson, *The Making of Man-Midwifery: Childbirth in England, 1660–1770* (Abingdon: Routledge, 2018).

Mary Fissell, *Vernacular Bodies: The Politics of Reproduction in Early Modern England* (Oxford: Oxford University Press, 2004).

Sara Read, *Maids, Wives, Widows: Exploring Early Modern Women's Lives, 1540–1740* (Havertown: Pen and Sword, 2015).

Doreen Evenden, *The Midwives of Seventeenth-Century London* (Cambridge: Cambridge University Press, 2006).

Insanity: The Mad and Their Doctors

Leonard Smith, *Private Madhouses in England, 1640–1815: Commercialised Care for the Insane* (Basingstoke: Palgrave Macmillan, 2020).

Catherine Arnold, *Bedlam: London and its Mad* (Basingstoke: Palgrave Macmillan, 2007).

Michael MacDonald and Terence Murphy, *Sleepless Souls: Suicide in Early Modern England* (Oxford: Clarendon Press, 1993).

Mary Ann Lund, *A User's Guide to Melancholy* (Cambridge: Cambridge University Press, 2021).

Those They Called Quacks: Unauthorised Healers

Peter Elmer, *The Miraculous Conformist: Valentine Greatrakes, the Body Politic, and the Politics of Healing in Restoration Britain* (Oxford: Oxford University Press, 2012).

Farah Karim-Cooper, *Cosmetics in Shakespearean and Renaissance Drama* (Edinburgh: Edinburgh University Press, 2006).

A. S. Hargreaves, *White As Whales Bone: Dental Services in Early Modern England* (Michigan: Northern Universities Press, 1998).

Roy Porter, *Quacks: Fakers and Charlatans in Medicine* (NPI Media Group, 2003).

Care-Takers and Criminals: Hospitals and the Welfare State

Paul Chambers, *Bedlam: London's Hospital for the Mad* (Ian Allen Publishing, 2009).

John Henderson, *The Renaissance Hospital: Healing the Body and Healing the Soul* (New Haven: Yale University Press, 2006).

Anne Borsay and Peter Shapely, eds, *Medicine, charity and mutual aid: the consumption of health and welfare in Britain, c.1550–1950* (Farnham: Ashgate, 2007).

Linda Woodbridge, *Vagrancy, Homelessness, and English Renaissance Literature* (University of Illinois, 2001).

Remaking the World: Experimenters

Lisa Jardine, *Ingenious Pursuits: Building the Scientific Revolution* (New York: Anchor Books, 2000).

Bill Bryson, ed., *Seeing Further: The Story of Science and the Royal Society* (London: Harper Press, 2010).

Holly Tucker, *Blood Work: A Tale of Medicine and Murder in the Scientific Revolution* (London: W. W. Norton, 2012).

Emily Cock, *Rhinoplasty and the nose in early modern British medicine and culture* (Manchester: Manchester University Press, 2019).

Thomas Wright, *Circulation: William Harvey's Revolutionary Idea* (London: Chatto and Windus, 2012).

Endnotes

All references to Shakespeare, unless otherwise stated, are taken from *Oxford Shakespeare: The Complete Works*, 2nd edition (Oxford: Oxford University Press, 2005).

Prologue

1 Graunt, J., *London's dreadful visitation, or, A collection of all the bills of mortality for this present year beginning the 20th of December, 1664, and ending the 19th of December following* (London, 1665), unnumbered page.

2 These are very moveable figures. At times, such as during some decades in the second half of the seventeenth century, average life expectancy dipped as low as thirty (probably because of poor harvests and outbreaks of plague). In some other years, it was in the fifties. Aristocrats were much more likely to live into old age than poor people, and male children usually did better than girls, perhaps because families funnelled more resources into their survival.

3 Wrigley, E.A., 'Mortality in Pre-Industrial England: The Example of Colyton, Devon, over Three Centuries', *Daedalus*, 1968, 97.2, pp. 546–80, esp. pp. 558–9.

4 Pepys, S., '16 September 1665', *The Diary of Samuel Pepys* [online], accessed 12 September 2023, www.pepysdiary.com.

Introduction: Know Thyself

1 Joannes, *The Englishmans Doctor; or. The School of Salerne* (London: R. Bradock, 1608), fol. C3r.

2 Though medical oaths are still often referred to as the 'Hippocratic oath', they are very different now to the original, which made students pledge that they would lend their teachers money whenever they required it, and would hold them in the same regard as their own parents.

3 Manning, J., *A New Booke, Intituled, I Am for You All, Complexions Castle* (Cambridge: John Legat, 1604), p. 11.

4 Manning, *A New Booke*, p. 28.

5 Joannes, *The Englishmans Doctor*, fol. C3v.

6 Wright, T., *The Passions of the Minde in Generall. Corrected, Enlarged, and with Sundry New Discourses Augmented* (London: Valentine Simms and Adam Islip, 1604), p. 45.

7 Wright, *The Passions of the Minde*, p. 47.

8 Joannes, *The Englishmans Doctor*, fol. A6v.

9 Elyot, T., *The Castel of Helth* (London: 1610), p. 27.

10 Wurtz, F. (trans. by R. Wurtz), *'The Children's Book'. Appended to An Experimental Treatise of Surgerie in Four Parts* [Translation of 'Practica de Wundartzney', 1563] (London: Gartrude Dawson, 1658), p. 51.

11 *An Account of the Causes of Some Particular Rebellious Distempers* (London: 1670), p. 22.

12 Manning, *A New Booke*, p. 5.

13 Hentzner, P. (trans. by H. Walpole), *Travels in England during the reign of Queen Elizabeth* (London: Edward Jeffery, 1797), p. 3.

14 Siraisi, N., *Medieval and Early Renaissance Medicine: An Introduction to Knowledge and Practice* (Chicago: University of Chicago Press, 1990), p. 40.

15 The flexor digitorum superficialis is responsible for flexing the fingers, while the tendon flexor pollicis longus flexes the thumb and contributes to moving the wrist.

16 *A Catalogue of All the Cheifest [sic] Rarities in the Publick Theater and Anatomie-Hall of the University of Leiden* (Leiden: Jacobus Voorn, 1683), fol. A2r.

17 *A Catalogue of All the Cheifest Rarities*, fols. 2v–3r.

18 Joffe, S. and V. Buchanan, 'The Vesalius "Epitome" of "De Humani Corporis Fabrica" of 1543 – A Worldwide Census with New Findings', *Medical Research Archives* [online], accessed 6 December 2022, https://esmed.org/MRA/mra/article/view/360, 2.1.

19 Unfortunately, the theatre was demolished in 1784, after several decades of redundancy. The theatre was designed by famed architect Inigo Jones, whose other designs included St James's Palace and Covent Garden square in London. He was also a producer of 'masques', stylised plays staged for the court with elaborate costumes and scenery.

20 Vesalius, A. (trans. by W. Richardson), *On the Fabric of the Human Body: A Translation of De Humanis Corporis Fabrica Libra Septem*, vol. 1 of 5 (San Francisco: Oxford Publishing, 1998), p. 63.

21 Languet, H. (trans. by C.D. O'Malley), 'Letter from Hubert Languet, diplomat, to Casper Peucer, physician, 1 January 1565' in O'Malley, C.D., *Andreas Vesalius of Brussels, 1514–1564* (Berkeley: University of California Press, 1964), p. 304. My italics.

22 Park, K., 'The Criminal and the Saintly Body', *Renaissance Quarterly*, 1994, 47.1, pp. 1–33, esp. p. 20.

23 Nashe, T., 'The Unfortunate Traveller' in J.B. Steane, ed., *The Unfortunate Traveller and Other Works* (London: Penguin, 1972), p. 349.

Health begins at home: domestic healers

1 Freke, E., 'Freke Papers. Vol. I. Commonplace Book of Elizabeth Freke, Begun Not Earlier than Sept. 1684 (See f. Ii) and Continued down to Feb. 1714', British Library Add MS 45718, p. 177.

2 Freke, E. (ed. R. Anselment), *The Remembrances of Elizabeth Freke 1671–1714* (Cambridge: Cambridge University Press, 2001), p. 37.

3 Freke, E., 'Freke Papers. Vol. II. Second Commonplace Book of Elizabeth Freke', British Library Add MS 45719, fol. 3r.

4 Freke, 'Freke Papers, Vol I', p. 5.

5 Cromwell, E., 'The Lady Elizabeth Cromwell to her Husband the Lord General at Edinburgh, December 27 1650' in T. Carlyle, ed., *Oliver Cromwell's Letters and Speeches*, vol. 1 of 2 (New York: Harper and Brothers Publishers, 1860), p. 517.

6 Pepys, S., '28 February 1664', *The Diary of Samuel Pepys* [online], accessed 12 September 2023, www.pepysdiary.com.

7 Amussen, S.D., '"Being stirred to much unquietness": Violence and Domestic Violence in Early Modern England', *Journal of Women's History*, 1994, 6.2, pp. 70–89.

8 This was the divorce of Lord Roos from his wife Anne on grounds of her infidelity. Roos obtained a separation in the ecclesiastical courts in 1666, then in 1667 persuaded Parliament to declare Anne's two sons illegitimate, and finally obtained a divorce in 1670.

9 Freke, 'Freke Papers, Vol. II', fol. 14v.

10 Freke, 'Freke Papers, Vol II', fol. 14v.

11 Cotta, J., *A Short Discouerie of Severall Sorts of Ignorant and Unconsiderate Practisers of Physicke in England* (London: W. Jones, 1619) pp. 21, 14.

12 Cotta, J., *The Triall of Witch-Craft* (London: George Purslow, 1616); Cotta, J., *The Infallible True and Assured Witch, or, The Second Edition of the Tryall of Witch-Craft* (London: I.E., 1625).

13 Markham, G., *The English House-Wife* (London: Nicholas Okes, 1631), pp. 4–5.

14 Pollock, L., 'Mildmay [Née Sharington], Grace, Lady Mildmay (c. 1552–1620), Memoirist and Medical Practitioner', *Oxford Dictionary of National Biography* [online], accessed 13 September 2024, https://www.oxforddnb.com/view/10.1093/ref:odnb/9780198614128.001.0001/odnb-9780198614128-e-45817.

15 Woolley, H. [misattributed], *The Gentlewomans Companion; or, A Guide to the Female Sex* (London: A. Maxwell, 1673), p. 180.

16 Verney, R., *Memoirs of the Verney Family*, vol. 1 of 2 (London: Longmans, Green and Co., 1904) pp. 375–6.

17 Verney, R., *Memoirs of the Verney Family*, vol. 2 of 2 (London: Longmans, Green and Co., 1904), p. 207.

18 Woolley, H., *The Queen-like Closet; or, Rich Cabinet* (London: R. Lowndes, 1670), pp. 11–12.

19 Pollock, 'Mildmay [Née Sharington], Grace, Lady Mildmay'.

20 Pollock, L., *With Faith and Physic: The Life of a Tudor Gentlewoman, Lady Grace Mildmay, 1552–1620* (Collins and Brown, 1993), pp. 140–1.

21 Okeover, E., 'Collection of Medical Receipts' (1675–1725) Wellcome Library MS 3712, p. 56. Elizabeth Okeover's book shows several different handwriting styles. Some of it looks to have been transcribed from a now-lost original, with other portions added later. It looks likely that the bulk of the initial recipes came from Elizabeth Okeover (1629–71), younger sister of Rowland Okeover, who was well known as a medical practitioner in the family's native Staffordshire. Those recipes stayed in the family and were put into this book, which was added to by Rowland's daughter, also named Elizabeth (1644–?). This Elizabeth later married Wolfstan Adderley in about 1666 – a surviving record shows that Elizabeth received a marriage portion of £1,400, plus various lands which were pledged to Wolfstan once Rowland and his wife died.

22 Freke, 'Freke Papers, Vol. I', p. 259.

23 Aske, K., 'Puppy Water, Beauty's Help', *Early Modern Medicine* [online], accessed 1 February 2023, https://earlymodernmedicine.com/puppy-water-beautys-help/.

24 Nagy, D.E., *Popular Medicine in Seventeenth-Century England* (Popular Press, 1988), p. 61.

25 Nagy, D.E., *Popular Medicine*, p. 60.

26 Hoby, M., '26 Aug 1601', *Diary of Lady Margaret Hoby*, British Library, MS Egerton 2614.

27 Pennell, S., 'Perfecting Practice? Women, Manuscript Recipes and Knowledge in Early Modern England' in V. Burke and J. Gibson, eds, *Early Modern Women's Manuscript Writing: Selected Papers from the Trinity/Trent Colloquium* (Farnham: Ashgate, 2004), pp. 237–58, esp. p. 240.

28 Nagy, D.E., *Popular Medicine*, p. 68.

29 Okeover, 'Collection of Medical Receipts', n.p; p. 79.

30 Partridge, J., *The Widdowes Treasure* (London: J. Roberts, 1595), fol. C2v.

31 Behn's life (1640–89) was a story to rival anything she ever wrote. Between writing plays, poems and novellas, she worked as a spy for

Charles II in Antwerp. She reported on the activities of anti-monarchy exiles, but repeatedly found herself in debt despite her many talents, and died in poverty.

32 Woolley, H., *The Gentlewomans Companion; or, A Guide to the Female Sex* (London: A. Maxwell, 1643), pp. 10–11.

33 Hobby, E., 'A Woman's Best Setting out Is Silence: the writings of Hannah Wolley' in G. MacLean, ed., *Culture and Society in the Stuart Restoration: Literature, Drama, History* (Cambridge: Cambridge University Press, 1995), pp.179–200.

34 Woolley, H., *The Accomplish'd Lady's Delight in Preserving, Physick, Beautifying, and Cookery* (London: B. Harris, 1675), title page.

35 Woolley, H., *The Queen-like Closet; or, Rich Cabinet* (London: R. Lowndes, 1670), pp. 307–11.

36 Woolley, H., *A Supplement to The Queen-like Closet, or, A Little of Everything* (London: R. Chiswel, 1680), fol. A2r.

37 Woolley, *A Supplement to The Queen-like Closet*, fols. 2v–3r.

38 Woolley, *The Queen-like Closet*, p. 17.

39 Zielińska, S., et al., 'Greater Celandine's Ups and Downs–21 Centuries of Medicinal Uses of *Chelidonium Majus* From the Viewpoint of Today's Pharmacology', *Frontiers in Pharmacology*, 2018, 9, pp. 1–29.

40 Pareek, A., et al., 'Feverfew (Tanacetum Parthenium L.): A systematic review', *Pharmacognosy Reviews*, 2011, 5.9, pp. 103–10.

41 Hamidpour, M., et al., 'Chemistry, Pharmacology, and Medicinal Property of Sage (*Salvia*) to Prevent and Cure Illnesses Such as Obesity, Diabetes, Depression, Dementia, Lupus, Autism, Heart Disease, and Cancer', *Journal of Traditional and Complementary Medicine*, 2014, 4.2, pp. 82–8.

42 Klunk, J., et al., 'Evolution of immune genes is associated with the Black Death', *Nature*, 2022, 611, pp. 312–9.

43 Woolley, *A Supplement to the Queen-like Closet*, p. 24.

44 Woolley, H., *The Ladies Directory in Choice Experiments & Curiosities of Preserving in Jellies, and Candying both Fruits & Flowers. also, an Excellent Way of Making Cakes, Comfits, and Rich Court-Perfumes. with Rarities of Many Precious Waters* (London: 1662), University of Glasgow Library, MS Wing 2nd edition, W3281.

45 Woolley, *The Gentlewomans Companion*, pp. 1–2.

46 Woolley, *A Supplement to The Queen-like Closet*, p. 132.

47 Woolley, *A Supplement to The Queen-like Closet*, p. 16.

48 Woolley, *A Supplement to The Queen-like Closet*, fol. A6r.

Cornering the Market: Physicians

1 Ward, J., *Notebook of John Ward, Vol. 7*, Folger Shakespeare Library [online], accessed 5 September 2016, https://digitalcollections.folger.edu/bib335079-486306, fol. 21r.

2 Ward, J., *Notebook of John Ward, Vol. 9*, Folger Shakespeare Library [online], accessed 5 September 2016, https://digitalcollections.folger.edu/bib335077-486302, fol. 150r.

3 Ward, *Notebook... Vol. 9*, n.p.

4 Ward, *Notebook... Vol. 9*, n.p.

5 Ward, *Notebook... Vol. 7*, fol. 25v.

6 Ward, J., *Notebook of John Ward, Vol. 4*, Folger Shakespeare Library [online], accessed 5 September 2016, https://digitalcollections.folger.edu/bib335072-486295, fols. 105v–6r.

7 Ward, J., *Notebook of John Ward, Vol. 11*, Folger Shakespeare Library [online], accessed 5 September 2016, https://digitalcollections.folger.edu/bib335079-486306, p. 601.

8 Ward, J., *Notebook of John Ward, Vol. 12*, Folger Shakespeare Library [online], accessed 5 September 2016, https://digitalcollections.folger.edu/bib335080-486308, fol. 40v.

9 Ward, J., *Notebook of John Ward, Vol. 14*, Folger Shakespeare Library [online], accessed 5 September 2016, https://digitalcollections.folger.edu/bib335081-486309, fol. 79v.

10 Ward, *Notebook... Vol. 11*, p. 615.

11 Nagy, D.E., *Popular Medicine*, p. 20.

12 Nagy, D.E., *Popular Medicine*, p. 23.

13 Hoyle, R.W., 'Famine as agricultural catastrophe: the crisis of 1622–4 in east Lancashire', *Economic History Review*, 2010, 63.4, pp. 974–1,002, esp. p. 978.

14 Nagy, D.E., *Popular Medicine*, p. 23.

15 King, E., *Sir Edmund King's Casebook (1676–1696)*, British Library, Sloane MS 1589.

16 Maynwaringe, E., *The Frequent, but Unsuspected Progress of Pains, Inflammations, Tumors, Apostems, Ulcers, Cancers, Gangrenes and Mortifications* (London: 1679), p. 36.

17 Harvey, G., *Little Venus Unmask'd* (1702), p. 65.

18 Read, A., *Chirurgorum Comes; or, The Whole Practice of Chirurgery* (London: E. Jones, 1687), p. 51.

19 Ward, *Notebook... Vol. 7*, fol. 36v.

20 Read, *Chirurgorum Comes*, p. 51; Denys, J.B., 'An Extract of a Letter of M. Denis Prof. of Philosophy and Mathematicks to M. * * * Touching the Transfusion of Blood, of April 2. 1667', *Philosophical Transactions of the Royal Society of London*, 1667, 2.25, pp. 453–5.

21 Willis, T., *Pharmaceuticae Rationalis* (London, 1679), p. 83.

22 Minderer, R., *Medicine Militaris: Or, a Body of Military Medicines Experimented* (London: 1674), p. 14.

23 Morton, R., *Phthisiologia, or, a treatise of consumptions* (London: Sam Smith and Benjamin Walford, 1674), p. 10.

24 Ward, J., *Notebook of John Ward, Vol. 10*, Folger Shakespeare Library [online], accessed 5 September 2016, https://digitalcollections.folger.edu/bib335078-486304, fol. 568.

25 Read, A., *The Works of That Famous Physician Dr. Alexander Read* (London: 1650), p. 173.

26 Platter, F., *Platerus Golden Practice of Physick* (London: Peter Cole, 1662), p. 40.

27 Tanner, J., *The Hidden Treasures of the Art of Physick* (1659), p. 116.

28 Cowper, S., *Diary, Vol. 1, 1700–1702*, Hertfordshire Records Office, MS D/EP F29, p.197.

29 McCray Beier, L., *Sufferers and Healers: The Experience of Illness in Seventeenth-Century England* (New York: Routledge, 2015), p. 77.

30 Wirsung, C. (trans. by J. Mosan), *Praxis Medicinae Universalis* (London: Edmund Bollifant, 1598), p. 572.

31 Thomson, G., *Galeno-pale, or, A chymical trial of the Galenists, that their dross in physick may be discovered* (London: R. Wood, 1665), p. 49.

32 Bonet, T., *A Guide to the Practical Physician Shewing from the most Approved Authors, both Ancient and Modern, the Truest and Safest Way of Curing all Diseases, Internal and External, Whether by Medicine, Surgery, Or Diet* (London: 1684), p. 817.

33 Barrough, P., *The Method of Physick* (London: 1583), p. 14.

34 Barrough, *The Method of Physick*, fol. A5r.

35 Ward, *Notebook... Vol. 4*, fol. 39v.

36 Cowper, *Diary, Vol. 1*, p. 169.

37 Ward, *Notebook... Vol. 4*, fol. 191.

38 It's not clear, but Henry's niece does not seem to have been his sister's daughter. Henry was the seventh son of his parents, so presumably 'Niece Ladde' was the daughter of one of his brothers. Henry's account of his relative's illness is recorded in letters to his close friend and pupil Lady Anne Conway. The letters, along with many other papers, were discovered in a chest in Ragley Hall in Warwickshire by the eighteenth-century prime minister Horace Walpole.

39 Read, *Chirurgorum Comes*, p. 147.

40 Turner, D., *De Morbis Cutanels: Diseases Incident to the Skin* (1714), p. 75.

41 Gabelkover, O., *The Boock of Physicke* (1599), p. 367.

42 Read, *The Works*, p. 175.

43 Bayfield, R., *Tractatus de Tumoribus Praeter Naturam, or, A Treatise of Preternatural Tumors* (1662), p. 189.

44 Lowe, P., *A Discourse of the Whole Art of Chyrurgerie. 3rd edition* (London: Thomas Purfoot, 1663), n.p.

45 More, H. and A. Conway (eds M. Nicolson and S. Hutton), *The Conway Letters: The Correspondence of Anne, Viscountess Conway, Henry More, and Their Friends, 1642–1684* (Oxford: Oxford University Press, 1992), p. 389.

46 Ward, *Notebook... Vol. 12*, p. 808.

47 More and Conway, *The Conway Letters*, p. 392.

48 More and Conway, *The Conway Letters*, pp. 393–4.

49 More and Conway, *The Conway Letters*, p. 398.

50 More and Conway, *The Conway Letters*, pp. 398–9.

51 More and Conway, *The Conway Letters*, p. 400.

52 More and Conway, *The Conway Letters*, p. 426.

53 More and Conway, *The Conway Letters*, p. 399.

54 Graunt, J., *London's Dreadful Visitation.*

55 Harvey, G., *A Discourse of the Plague* (London: Nathaniel Brooke, 1665), p. 9.

56 *A True Bill of the Whole Number That Hath Died* (London: J. Roberts, 1603).

57 *A Generall or Great Bill for This Yeere* (London: William Standsby, 1625).

58 Pepys, S., '7 June, 1665', *The Diary of Samuel Pepys* [online], accessed 12 September 2023, www.pepysdiary.com.

59 Pepys, '16 August, 1665'.

60 Royal College of Physicians of London, *Certain necessary directions as well for the cure of the plague, as for preventing the infection: with many easie medicines of small charge, very profitable to His Majesties subjects* (London: Christopher Barker, 1665), p. 1.

61 Thomson, *Galeno-Pale*, p. 53.

62 Defoe was only five years old at the time of the plague, so much of his 'journal' was based on the recollections of his uncle Henry, who lived in Aldgate. D. Defoe (ed. Cynthia Wall), *A Journal of the Plague Year*, (London: Penguin, 2003), pp. 66–67.

63 This fear was later exemplified in a 1782 engraving depicting William Hunter's anatomical museum in Edinburgh. At the day of resurrection, the inhabitants of the museum are left squabbling over whose body parts belong to whom, calling out 'Where's my head?' and 'Damn you Sir that's *my* leg'. (*William Hunter (1718–1783) in his museum in Windmill Street on the day of resurrection, surrounded by skeletons and bodies, some of whom are searching for their missing parts.* Engraving, 1782. Wellcome Collection 25435i.)

64 Thomson, G., *Loimotomia, or, The Pest Anatomized* (London: Nathaniel Crouch, 1666), p. 3.

65 Thomson, *Loimotomia*, p. 109.

66 *By the King, a Proclamation for a Generall Fast throughout This Realm of England* (London: John Bill and Christopher Barker, 1665).

67 Kemp, W., *A Brief Treatise of the Nature, Causes, Signes, Preservation from, and Cure of the Pestilence* (London: 1665), p. 17.

68 Harvey, *A Discourse of the Plague*, p. 13.

69 Harvey, *A Discourse of the Plague*, pp. 15–16.
70 Kemp, *A Brief Treatise*, p. 16.
71 Thomson, *Loimotomia*, pp. 8–9, 17–18.
72 Thomson, *Loimotomia*, p. 35.
73 Thomson, *Loimotomia*, pp. 122, 131.
74 Thomson, *Loimotomia*, pp. 131–2.
75 Thomson, *Loimotomia*, p. 146.
76 Thomson, *Loimotomia*, p. 147.
77 Thomson, *Loimotomia*, pp. 150, 152–3, 155.
78 Thomson, *Loimotomia*, p. 157.
79 Royal College of Physicians of London, *Certain Necessary Directions*, p. 20.
80 Thomson, *Loimotomia*, pp. 160–1.
81 Thomson, *Loimotomia*, p. 170.
82 Thomson, *Loimologia*, p. 2.
83 Thomson, *Loimologia*, pp. 10–11.
84 Elmer, P., 'Society of Chemical Physicians (Act. 1665–1666)', *Oxford Dictionary of National Biography* [online], accessed 6 June 2022, https://www.oxforddnb.com/display/10.1093/ref:odnb/9780198614128.001.0001/odnb-9780198614128-e-107538.

Right place, right time: surgeons

1 Ward, *Notebook… Vol. 4*, fol. 99r.
2 Dionis, P., *A course of chirurgical operations, demonstrated in the Royal Garden at Paris* (London: 1710), pp. 254–5.
3 Bonet, *A Guide to the Practical Physician*, p. 62.
4 Dionis, *Chirurgical operations*, p. 255.
5 Read, A., *A Treatise of the First Part of Chirurgerie* (London: John Haviland, 1638) (Amsterdam; Norwood, NJ: Theatrum Orbis Terrarum, Ltd, and Walter J. Johnson, Inc., 1976) (Printed Facsimile Edition from STC 20786), p. 43.
6 Woodall, J., 'A Treatise of Gangraena', appended to *The surgeons mate, or, Military & domestique surgery: discovering faithfully & plainly [the] method*

and order of [the] surgeons chest, [the] uses of the instruments, the vertues and operations of [the] medicines, with [the] exact cures of wounds made by gun-shott; and otherwise (London: Rob, Young, 1639), p. 189.

7 Ward, J., *Notebook of John Ward, Vol. 8*, Folger Shakespeare Library [online], accessed 5 September 2016, https://digitalcollections.folger.edu/bib335076-486301, fol. 117v.

8 Ancillon, C., *Italian Love: Or, Eunuchism Displayed. 2nd edition* (London: E. Curll, 1740), p. 16.

9 James, R., 'Amputation' in *A Medicinal Dictionary*, vol. 1 of 3 (London: T. Osborne and J. Roberts,1743–5).

10 Bonet, *A Guide to the Practical Physician*, p. 62.

11 Turner, D., *An Appendix to Dr. Turner's Art of Surgery* (London: John Clark, 1725).

12 Woodall, 'A Treatise of Gangraena', p. 388; Ward, *Notebook…Vol. 8*, fol. 32r.

13 Dionis, *Chirurgical Operations*, p. 256.

14 *London Evening Post*, Issue 27 (London: 8–10 February 1728); *Daily Courant*, Issue 5347 (London: 28 May 1733); *Daily Gazetteer*, Issue 919 (London, 14 June 1738).

15 Lowe, P., *A Discourse of the Whole Art of Chyrurgerie. 3rd edition* (London: Thomas Purfoot, 1663), n.p.

16 Thomas, D.P., 'Thomas Vicary, Barber-Surgeon', *Journal of Medical Biography*, 2006, 14.2, pp. 84–9.

17 In later years, Vicary was rewarded richly for his service, being given land and a house at Boxley in Kent. In 1546 he was also made the resident superintendent of St Bartholomew's Hospital.

18 Clowes, W., *A Profitable and Necessarie Booke of Obseruations, for All Those That Are Burned with the Flame of Gun Powder, &c.* (London: Edmund Bollifant, 1596).

19 Chamberland, C., 'Partners and Practitioners: Women and the Management of Surgical Households in London 1570–1640', *Social History of Medicine*, 2011, 24.3, pp. 553–69, esp. p. 559.

20 Fabry, W., *Lithotomia vesicae: that is, An accurate description of the stone in the bladder shewing the causes and pathognomicall signes thereof, and chiefely of the method whereby it is to be artificially taken out both of men and women, by section* (London: John Norton, 1640), p. 93.

21 Josselin, R. (ed. A. Macfarlane), *The Diary of Ralph Josselin, 1616–1683* (Oxford: Oxford University Press, 1976), p. 468; Evenden, D.A., 'Gender differences in the licensing and practice of female and male surgeons in early modern England', *Medical History*, 1998, 42.2, pp. 194–216, esp. pp. 206, 208.

22 In 1571, this structure was expanded so that up to twenty people at once could be hanged, watched by eager crowds.

23 Fabry, *Lithotomia vesicae*, p. 49.

24 Wurtz, *'The Children's Book'*, p. 3.

25 Carlton, C., *This Seat of Mars: War and the British Isles, 1485–1746* (New Haven: Yale University Press, 2011), p. 37.

26 Morrow, C., 'Corporate Nationalism in Thomas Dekker's *The Shoemaker's Holiday*', *SEL Studies in English Literature 1500–1900*, 2014, 54.2, pp. 423–54, esp. p. 427.

27 'News from the Camp, on Black Heath, or The Noble Souldiers Resolution: Expressing His Heroick Courage to serve his King, and Country, to his utmost Ability, Through all Dangers and Exigences', *English Broadside Ballad Archive* [online], accessed 7 August 2022, https://ebba.english.ucsb.edu/ballad/36927/image.

28 Barwick, H., *A Breefe Discourse, Concerning the Force and Effect of All Manuall Weapons of Fire* (London: E. Allde, 1592), p. 14.

29 Barwick, *A Breefe Discourse*, p. 11.

30 Nichols, E., *The Dolphins Danger and Deliverance* (London: Henry Gosson, 1617), fol. C4v.

31 Barry, J., 'John Houghton and Medical Practice in London c. 1700', *Bulletin of the History of Medicine*, 2018, 92.4, pp. 575–603.

32 Moyle, J., *Abstractum Chirurgiae Marinae., or, An Abstract of Sea Chirurgery* (London: J. Richardson, 1686), p. 22.

33 Bynum, W.F., *Science and the Practice of Medicine in the Nineteenth Century* (New York: Cambridge University Press, 1994), pp. 26–7.

34 Paré, A. (trans. by Th: Johnson), *The Workes of That Famous Chirurgion Ambrose Parey* (London: Th: Cotes and R. Young, 1634), p. 408.

35 Paré, A. (trans. by R.W. Linker and N. Womack), *Ten Books of Surgery with the Magazine of the Instruments Necessary for It* (Georgia, US: University of Georgia Press, 2010), p. 44.

36 Paré, *Ten Books*, p. 409.

37 Paré, *Ten Books*, p. 409.
38 Paré, *Ten Books*, p. 409.
39 Paré, *The Workes*, p. 1,133.
40 Paré, *The Workes*, p. 1,149.
41 O'Malley, C.D., *Andreas Vesalius of Brussels, 1514–1564* (Berkeley: University of California Press, 1964), pp. 286–8.
42 Paré, *Ten Books*, p. 73.
43 Paré, *Ten Books*, p. 85.
44 Paré, *The Workes*, p. 874.
45 Paré, *The Workes*, p. 874.
46 Paré, *The Workes*, p. 871.
47 Paré, *The Workes*, pp. 876–7.
48 Paré, *The Workes*, p. 144.
49 Paré, *The Workes*, p. 880.
50 Paré, *The Workes*, p. 880.
51 Woodall, 'A Treatise of Gangraena', p. 386.
52 Woodall, *The surgeon's mate*, p. 156.
53 Paré, *The Workes*, p. 458.
54 de la Vauguion, M., *A Compleat Body of Chirurgical Operations* (London: 1699), p. 291.
55 'Manuscript lecture notes, seventeenth century', Wellcome Library, MS.MSL.3.
56 *London Gazette* (London, England), 29 November 1705 – 3 December, 1705; Issue 4180, p.2. Read also cautioned readers against 'circumferaneous Pretenders' who he said travelled the country pretending to be one of his servants and doing cowboy operations.
57 *Weekly Journal or British Gazetteer* (London, England), Saturday, 9 October, 1725; Issue 24.
58 Middlesex Sessions: Sessions Papers – Justices' Working Documents. October 1727. *London Lives* ref: LMSMPS502470026 (londonlives.org). My italics.
59 Thorpe, J., *Mr. Penny's dishonourable breach of trust exemplified, at an amputation of a thigh: being called as an assistant to that operation* (London: 1721).
60 Fabry, *Lithotomia vesicae*, pp. 29–30.
61 Ward, *Notebook…vol. 12*, fol. 777.

Medicine for the masses: apothecaries

1 Wallis, P., 'Apothecaries and Medicines in Early Modern London', in L. Hill Curth, ed., *From Physick to Pharmacology: Five Hundred Years of British Drug Retailing* (Farnham: Ashgate, 2006), pp. 13–27, esp. p. 18.
2 Bacon, F., *A True and Historical Relation of the Poysoning of Sir Thomas Overbury* (London: T.M. & A.C., 1651), p. 21.
3 Bacon, *A True and Historical Relation*, p. 26.
4 Bacon, *A True and Historical Relation*, p. 4.
5 Bacon, *A True and Historical Relation*, p.13.
6 'A Page, a Knight', British Library, MS Egerton 2230, fol. 70v.
7 Bacon, *A True and Historical Relation*, p. 97.
8 Bacon, *A True and Historical Relation*, p. 98.
9 Bacon, *A True and Historical Relation*, p. 65.
10 Somerset, A., *Unnatural Murder: Poison In The Court Of James I* (London: Hachette UK, 2021), p. 213.
11 In Anne Somerset's excellent book on the Overbury affair, she makes a convincing case that Overbury was as likely to have died from the misguided attentions of King James's physician, Théodore Turquet de Mayerne, as from the more sinister machinations of Howard and Turner. Among de Mayerne's treatments was bleeding Overbury over a sustained period by placing dried peas in an incision so that it would not heal – known as an 'issue' (Somerset, *Unnatural Murder*, pp. 228–9). Either way, though, the public perception was that Overbury had been killed by poison, which effected distrust in medicine and apothecaries.
12 *James Franklin, a Kentish Man of Maidstone* (London: J.T., 1615).
13 Bacon, *A True and Historical Relation*, p. 51.
14 Cook, H.J., *Trials of an Ordinary Doctor: Joannes Groenevelt in Seventeenth-Century London* (Baltimore: Johns Hopkins University Press, 1994), p. 6.
15 Bacon, *A True and Historical Relation*, p. 18.
16 Woodall, *The surgeon's mate*, p.229.
17 Cowper, S., 'May 16, 1704', *Diary, Vol. 1, 1700–1702*, Hertfordshire Records Office, MS D/EP F29 p. 219.
18 Salmon, W., *Ars Chirurgica* (London: J. Dawks, 1698), p. 235.
19 The strong effects of mercury mean that it was used in traditional Chinese, Ayurvedic and Unani (Persian) medicine for many centuries

before it became popular in Europe. It remained in use as a cure for syphilis well into the nineteenth century, and modern scholars are divided as to whether mercury ameliorated signs of infection or made them worse.

20 Culpeper, N., *Culpeper's School of Physick. Or The Experimental Practice of the Whole Art* (London: N. Brook, 1659), fol. B4r.

21 Culpeper, N., *Culpeper's School*, fol. B8r.

22 The identity of the lady is not known, but she had an estate of £2,000, plus £500 a year, making her financially out of Culpeper's league (Woolley, *The Herbalist*, pp. 30–1).

23 Woolley, *The Herbalist* p. 213.

24 Woolley, *The Herbalist*, p. 213.

25 Kassell, L., *Medicine and Magic in Elizabethan London: Simon Forman: Astrologer, Alchemist, and Physician* (Oxford: Oxford University Press, 2007), p. 118.

26 Culpeper, N., *The English Physitian, or An Astrologo-Physical Discourse of the Vulgar Herbs of This Nation* (London: Peter Cole, 1652), n.p.

27 Nearly a decade later, Lilly and other astrologers would predict the solar eclipse of Monday, 29 March 1652. Nicholas rushed to publish two works claiming that the event heralded 'the Ruine of Monarchy throughout Europe', but in the event, neither international politics nor the brightness of the sky over London looked much different from normal, leading to much public ridicule.

28 Woolley, *The Herbalist*, p. 126.

29 Culpeper, N., *Pharmacopoeia Londinensis, or the London Dispensatory* (London: Peter Cole, 1653), pp. 35–6.

30 Culpeper, *Pharmacopoeia*, p. 67.

31 Prevorius, J. (trans. by N. Culpeper), *Medicaments for the Poor; Or, Physick for the Common People* (London: Peter Cole, 1656), fol. B4r.

32 Culpeper, N., *A Physical Directory, or, A Translation of the Dispensatory Made by the Colledge of Physitians of London, and by Them Imposed upon All the Apothecaries of England to Make up Their Medicines By*, 2nd edition (London: Peter Cole, 1650), fol. B2r.

33 Furdell, E.L., '"Reported to Be Distracted": The Suicide of Puritan Entrepreneur Peter Cole', *Historian*, 2004, 66.4, pp. 772–92.

34 Culpeper, *The English Physitian*, p. 5.
35 Culpeper, *The English Physitian*, p. 17.
36 Chickweed continues to be used widely in herbal medicine as an anti-inflammatory, though there is little solid scientific evidence for its efficacy.
37 Culpeper, *A Physical Directory*, fol. B2r.
38 Culpeper, *Culpeper's School*, fol. B8v.
39 Culpeper, N., *Mr Culpeper's Ghost, Giving Seasonable Advice to the Lovers of His Writing. Before Which Is Prefixed, Mris. Culpepers Epistle in Vindication of Her Husband's Reputation* (London: Peter Cole, 1656), fols. 2v–3r.
40 Ward, *Notebook… Vol. 1*, pp. 106–7.
41 Culpeper, *Mr Culpeper's Ghost*, p. 4.
42 Culpeper, *Mr Culpeper's Ghost*, p.11.
43 Furdell, 'Reported to Be Distracted', p. 791.
44 Whittet, T.D., *The Apothecaries in the Great Plague of London, 1665* (London: Society of Apothecaries, 1965), p. 3.
45 Boghurst, W., *Loimographia* (London: Shaw And Sons, 1894), pp. 30–1.
46 Whittet, *The Apothecaries in the Great Plague of London.*

In the Beginning: Midwives

1 Freke, 'Freke Papers. Vol. II', p. 4.
2 Aristocratic women would often give birth many more times, because breastfeeding had a contraceptive effect and wealthy women tended not to breastfeed their own children, instead sending them out to wet-nurses. (Read, *Maids, Wives, Widows*, p. 98).
3 Lewis, J., '"Tis a Misfortune to Be a Great Ladie": Maternal Mortality in the British Aristocracy, 1558–1959', *Journal of British Studies*, 1998, 37.1, pp. 26–53.
4 Evenden-Nagy, D., *Seventeenth-century London midwives: Their training, licensing and social profile* [thesis], McMaster University [online], accessed 7 July 2021, https://www.proquest.com/docview/230817240, pp. 152–4.
5 Wellcome Library MS. 3544/2.
6 Evenden-Nagy, *Seventeenth-century London midwives*, p. 186.

7 Wurtz, *'The Children's Book'*, pp. 341–2.
8 Evenden-Nagy, '*Seventheenth-century London midwives*', p. 121.
9 Sharp, J., *The Midwives Book* (London: Simon Miller, 1671), p. 33.
10 Sharp, *The Midwives Book*, pp. 43–4.
11 Ward, *Notebook... Vol. 12*, p. 896.
12 Sharp, *The Midwives Book*, p. 40.
13 Sharp, *The Midwives Book*, pp. 40–1.
14 Sharp, *The Midwives Book*, p. 129.
15 Bartholin, T., *Bartholinus Anatomy* (London: John Streater, 1668), p. 89.
16 Sadler, J., *The Sicke Woman's Private Looking-Glasse. Wherein Methodically Are Handled All Uterine Affects or Diseases Arising from the Womb* (London: Anne Griffin, 1636), p. 61.
17 Sharp, *The Midwives Book*, p. 180.
18 Sharp, *The Midwives Book*, p. 103.
19 Sharp, *The Midwives Book*, p. 103.
20 Sharp, *The Midwives Book*, p. 104.
21 Thornton, A. (ed. C. Jackson), *The Autobiography of Mrs. Alice Thornton* (Publications of the Surtees Society. Vol. 62.), pp. 164–5.
22 Pollock, L., 'Embarking on a rough passage: the experience of pregnancy in early-modern society' in V. Fildes, ed., *Women as Mothers in Pre-Industrial England* (London: Routledge, 1993), pp. 39–67, esp. p. 55.
23 Pollock, 'Embarking on a rough passage', p. 55.
24 Pollock, 'Embarking on a rough passage', p. 39.
25 Pollock, 'Embarking on a rough passage', p. 46.
26 Freke, 'Freke Papers. Vol. II', fol. 47r.
27 Freke, 'Freke Papers. Vol. II', fol. 48r–v.
28 Thornton, *The Autobiography of Mrs. Alice Thornton*, p. 87.
29 Thornton, *The Autobiography of Mrs. Alice Thornton*, p. 88.
30 Sharp, *The Midwives Book*, pp. 181–2; 'Whitbread Family Receipt Book', Wellcome Library MS 8745, p. 19.
31 Preparing the linens was not only for practical purposes. If she suffered a late miscarriage or the baby was stillborn without witnesses, the mother could shield herself from accusations of abortion or infanticide by using the linens as evidence that she had been preparing for and looking forward to a live baby.

32 Josselin, R. (ed. A. Macfarlane), *The Diary of Ralph Josselin 1616–1683* (Oxford: Oxford University Press, 1976), p. 50.

33 Fissell, M.E., 'The Politics of Reproduction in the English Reformation', *Representations*, 2004, 87.1, pp. 43–81, esp. p. 45.

34 Culpeper, N., *A directory for midwives: or, A guide for women, in their conception, bearing, and suckling their children* (Edinburgh: George Swintoun and James Glen, 1655), p. 116.

35 Lupton, T., *A Thousand Notable Things, of Sundry Sortes* (London: John Charlewood, 1579), vervain: p. 62; dittany: p. 12; theriac: p. 146; snake skin: p. 86; gourd: p. 93; oakfern: p. 193; sage: p. 168; mugwort: p. 181.

36 Red coral was believed to be powerful in protecting children from all kinds of harm, from epilepsy to the evil eye. It was made into rattles, teething necklaces and amulets, and given in medicines. Nicholas Culpeper advised in his *Physical Directory*, 'If ten grains of red Coral be given to a child in a little breast-milk so soon as it is born, before it take any other food, it will never have the falling sickness [epilepsy], nor convulsions', (p. 35).

37 Thornton, *The Autobiography of Mrs. Alice Thornton*, p. 95.

38 Sharp, *The Midwives Book*, pp. 201–2.

39 Wilson, A., *The Making of Man-Midwifery: Childbirth in England, 1660–1770* (Boston: Harvard University Press, 1995), p. 53.

40 Chamberlen, P., *A Voice in Rhama* (London: William Bentley, 1647), fol. A4r.

41 'Rhama' here refers to Jeremiah 31:15, 'a voice was heard in Ramah, lamentation and bitter weeping; Rachel, weeping for her children, refused to be comforted for her children, because they were no more'. Rachel represents the grief of mothers who have lost their children, so this title was apt to Chamberlen's subject, and a good way for him to display his scriptural knowledge.

42 Chamberlen, *A Voice in Rhama*, n.p.

43 Mauriceau, F., *The Accomplisht Midwife, Treating of the Diseases of Women with Child, and in Child-Bed* (London: J. Darby, 1673), fol. A2r.

44 Dunn, P.M., 'The Chamberlen Family (1560–1728) and Obstetric Forceps', *Archives of Disease in Childhood – Fetal and Neonatal Edition*, 1999, 81.3, p. 233.
45 King, H., 'Chamberlen, Hugh, the Elder (b. 1630x34, d. after 1720), Physician and Economist', *Oxford Dictionary of National Biography* [online], accessed 9 January 2022, https://www.oxforddnb.com.
46 Culpeper, *A Directory for Midwives*, p. 120.
47 Culpeper, *A Directory for Midwives*, p. 121.
48 Sharp, *The Midwives Book*, p. 217.
49 Gray, J.M., ed., *Memoirs of the life of Sir John Clerk of Penicuik* (Edinburgh: Scottish Historical Society, 1892), pp. 39–40.
50 Sharp, *The Midwives Book*, p. 229.
51 Cellier, E., 'A Scheme for the Foundation of a Royal Hospital', *A Fourth Collection of Scarce and Valuable Tracts* (London: 1751), fol. Ii3r.
52 *Modesty Triumphing over Impudence, or, Some Notes upon a Late Romance Published by Elizabeth Cellier, Midwife and Lady Errant* (London: Jonathan Wilkins, 1680), p. 17.
53 Cellier, E., *To Dr. —— an Answer to His Queries Concerning the Colledg of Midwives* (London, s.n., 1688).

Insanity: the Mad and Their Doctors

1 Burton, R., *The Anatomy of Melancholy* (Oxford: John Lichfield and James Short, 1621), p. 3.
2 Burton, *The Anatomy of Melancholy*, p. 371.
3 Burton, *The Anatomy of Melancholy*, p. 6.
4 Lund, M.A., *A User's Guide to Melancholy* (Cambridge: Cambridge University Press, 2021), p. 72.
5 Irish, D., *Levamen Infirmi* (London: 1700), p. 45.
6 Newton, J., 'They that have any friends distracted or melancholy are desired to accept this' (London: 1675?).
7 Blackmore, R., *A Treatise of the Spleen and Vapours: Or, Hypocondriacal and Hysterical Affections. With Three Discourses on the Nature and Cure of the Cholick, Melancholy, and Palsies*, 2nd edition (London: J. Pemberton,

1726), p. 161; Robinson, R., *A new system of the spleen, vapours, and hypochondriack melancholy* (London: Samuel Aris, 1729), p. 230.

8 Burton, *The Anatomy of Melancholy*, p. 16.

9 Burton, *The Anatomy of Melancholy*, p. 233.

10 Robinson, *A New System*, p. 229.

11 Robinson, *A New System*, p. 234.

12 In the seventeenth century, fewer than three hundred Londoners each year died of 'madness', out of a population of between 200,000 and 550,000. Boulton, J. and Black, J., "Those, that die by reason of their madness': dying insane in London, 1629–1830' *History of Psychiatry*, 23.1 (2012), pp.27-39, 29.

13 Many suicides were tragically young. The records of the King's Bench Court document 13,698 suicides between 1485 and 1714. Only 1,098 of those had their age recorded, but of that number, a third were aged under twenty-one. (Murphy, T.R., '"Woful Childe of Parents Rage": Suicide of Children and Adolescents in Early Modern England, 1507–1710', *Sixteenth Century Journal*, 1986, 17.3, pp. 259–70, esp. p. 262.)

14 Zell, M., 'Suicide in Pre-Industrial England', *Social History*, 1986, 11.3, p. 309.

15 Langley, E.F., *Narcissism and Suicide in Shakespeare and His Contemporaries* (Oxford: Oxford University Press, 2009), pp. 212–3.

16 Sym's was not the first text *written* on this topic; in 1608, John Donne wrote *Biathanatos*, a long and complex consideration of the ethics of self-killing. He concluded that suicide was not always a sin, but recognised that this opinion was controversial, and sent a copy to his friend Sir Robert Ker with the instruction to 'publish it not, but yet burn it not; and between those, do what you will with it'. (Donne, *Letters to Severall Persons of Honour Written by John Donne*, p. 22). *Biathanatos* was finally published in 1644, thirteen years after the author's death.

17 Sym, J., *Lifes Preservative against Self-Killing. Or, An Useful Treatise Concerning Life and Self-Murder* (London: 1637), pp. 2–3.

18 Baxter, R., *Preservatives against Melancholy and Over-Much Sorrow; or the Cure of Both* (London: Joseph Marshall, 1716), pp. 22–3.

19 Burton, *The Anatomy of Melancholy*, pp. 276–7.

20 Tryon, T., *A treatise of dreams & visions wherein the causes, natures, and uses, of nocturnal representations, and the communications both of good and evil angels,*

as also departed souls, to mankind are theosophically unfolded (London: 1689), p. 261.

21 Tryon, *A treatise of dreams*, p. 266.

22 Pettigrew, T., S. Pettigrew and J. Bailly, *The Major Works of John Cotta: The Short Discovery (1612) and the Trial of Witchcraft (1616)* (Boston: Brill, 2018), p. 160.

23 Cotta, J., *The Triall of Witch-Craft* (London: George Purslow, 1616).

24 Pettigrew et al., *The Major Works of John Cotta*, pp. 155, 157.

25 Thwaite, A., 'What is a "witch-bottle"? Assembling the textual evidence from early modern England', *Magic, Ritual, and Witchcraft*, 2020, 15.2, pp. 227–51.

26 Allen, H., *A Narrative of God's Gracious Dealings with That Choice Christian Mrs. Hannah Allen (Afterwards Married to Mr. Hatt)* (London: John Wallis, 1683), p. 4.

27 Allen, *A Narrative*, p. 18.

28 Allen, *A Narrative*, p. 21.

29 Allen, *A Narrative*, p. 28.

30 Allen, *A Narrative*, p. 46.

31 Allen, *A Narrative*, p. 36.

32 Scot, R., *Scots Discovery of Witchcraft* (London: 1654), fol. B2v.

33 Allen, *A Narrative*, p. 29.

34 Allen, *A Narrative*, pp. 40–1, 44.

35 Allen, *A Narrative*, p. 70.

36 Allen, *A Narrative*, p.72.

37 Allen, *A Narrative*, p.73.

38 Rivière's book was a particular favourite of John Ward, who cites it many times in his diaries.

39 Rivière, L. (trans. by N. Culpeper, A. Cole and W. Rowland), *The Practice of Physick, in Seventeen Several Books* (London: 1655), p. 49.

40 Nashe, T., *The Terrors of the Night or, A Discourse of Apparitions*, University of Michigan [online], accessed 17 June 2018, http://name.umdl.umich.edu/A08014.0001.001.

41 Burton, *The Anatomy of Melancholy*, pp. 80–117.

42 Burton, *The Anatomy of Melancholy*, pp. 192–4, 167.

43 Lund, *A User's Guide to Melancholy*, p. 168.

44 Ward, *Notebook... Vol. 7*', fol. 15r.

45 Bright, T., *A treatise of melancholie, Containing the causes thereof, & reasons of the strange effects it worketh in our minds and bodies: with the physicke cure, and spirituall consolation for such as have thereto adjoyned an afflicted conscience* (London: Thomas Vautrollier, 1586), p. 240.

46 Burton, *The Anatomy of Melancholy*, p. 363.

47 Burton, *The Anatomy of Melancholy*, pp. 636–7.

48 Bragg Ewald Jr., W., *Rogues Royalty And Reporters: The Age Of Queen Anne Through Its Newspapers* (Boston: Houghton Mifflin, 1954), p. 93.

49 Burton, *The Anatomy of Melancholy*, p. 460.

50 Bright, *A Treatise of Melancholy*, pp. 264–5; Burton, *The Anatomy of Melancholy*, p. 440.

51 'Digestive Powder', *Post Man and the Historical Account*, Issue 950, 31 March 1702–2 April 1702.

52 Fitzherbert, D., *Meditations and Diary*, Bodleian Library MS. e Mus. 169, f. 10r.

53 Hodgkin, K., 'Fitzherbert, Dionys (c.1580–c.1641), Writer on Madness', *Oxford Dictionary of National Biography* [online], accessed 7 October 2023, https://www.oxforddnb.com/view/10.1093/odnb/9780198614128.001.0001/odnb-9780198614128-e-112759.

54 Smith, L., *Private Madhouses in England, 1640–1815: Commercialised Care for the Insane* (London: Palgrave Macmillan, 2020), p. 22.

55 Smith, *Private Madhouses*, p. 22.

56 Vicars, J., *A Looking-Glasse for Malignants, or, Gods Hand against God-Haters* (London: Iohn Rothwell, 1643), pp. 15, 22.

57 Irish, *Levamen Infirmi*, p. 53.

58 Newton, 'They that have any friends'.

59 Mason, A., 'The Reverend John Ashburne (c.1611–61) and the Origins of the Private Madhouse System', *History of Psychiatry*, 1994, 5.19, pp. 321–45, esp. p. 336.

60 Hunter, R. and I. MacAlpine, 'The Reverend John Ashbourne (C.1611–61) And The Origins Of The Private Madhouse System', *The British Medical Journal*, 1972, 2.5812, pp. 513–5.

61 Mason, 'The Reverend John Ashburne', p. 331.

62 Hunter and MacAlpine, 'The Reverend John Ashbourne', p. 514.

63 Mason, 'The Reverend John Ashburne', p. 341.

64 Westover, J. (transcribed by W.G. Hall), 'The Casebook of John Westover of Wedmore, Surgeon, 1686–1700', Somerset Heritage Centre, DD/X/HALW 4, December 1992, Revised July 1999, fol. 1v.
65 Westover, 'Casebook', fol. 16v.
66 Westover, 'Casebook', fols. 25v, 36v.
67 Westover, 'Casebook', fol. 39r.
68 Peterson, D., ed., *A Mad People's History of Madness* (Pittsburgh: University of Pittsburgh Press, 1982), pp. 26–38.
69 Westover, 'Casebook', fol. 209r.
70 Westover, 'Casebook', fol. 180v.
71 Westover, 'Casebook, fol. 8.
72 MacDonald, M. and T.R. Murphy, *Sleepless Souls: Suicide in Early Modern England* (Oxford: Clarendon Press, 1993), p. 124.
73 Smith, *Private Madhouses*, p. 26.
74 Carkesse, J., *Lucida Intervalla, Containing Divers Miscellaneous Poems, Written at Finsbury and Bethlem* (London: 1679), p. 31.
75 Smith, *Private Madhouses*, p. 27.

Those they called quacks: unauthorised healers

1 Pettigrew et al., *The Major Works of John Cotta*, pp. 34–5.
2 Pettigrew et al., *The Major Works of John Cotta*, p. 119.
3 In 1620–21, Cotta became embroiled in a case against 'Mistress Moyle' for allegedly fatally poisoning Sir Euseby Andrews under the guise of giving him medicine. Despite Cotta's testimony that the body showed signs of poisoning, Moyle was found innocent and a counter-claim made (verdict unknown) of conspiracy to falsely accuse her of murder. (John Taylor, *Tracts Rare and Curious Reprints, Ms. Etc: Relating to Northamptonshire. 2nd Series* (Elliot Stock, 1881), n.p.)
4 Pettigrew et al., *The Major Works of John Cotta*, p. 196.
5 Pettigrew et al., *The Major Works of John Cotta*, p. 217.
6 Pettigrew et al., *The Major Works of John Cotta*, p. 217.
7 Jonson, B. (ed. B. Parker), *Volpone* (Manchester: Manchester University Press, 1999), 2.2., pp. 108–13.

8 Moulton, T., *The Compleat Bone-Setter* (London: J.C., 1656), pp. 8–9.

9 *Post Man and the Historical Account* (London, England), Issue 11050, 5 May 1715–7 May 1715.

10 Cook, H.J., *Trials of an Ordinary Doctor: Joannes Groenevelt in Seventeenth-Century London* (Baltimore: Johns Hopkins University Press, 1994).

11 *Newes Published for Satisfaction and Information of the People*, 3 November 1664.

12 Raymond, J., 'Nedham [Needham], Marchamont (bap. 1620, d. 1678), journalist and pamphleteer', *Oxford Dictionary of National Biography* [online], accessed 16 September 2024, https://www.oxforddnb.com/view/10.1093/ref:odnb/9780198614128.001.0001/odnb-9780198614128-e-19847.

13 'Aulicus His Hue and Cry' (London: 1645), p. 1.

14 'Advertisements and Notices', *London Post with Intelligence Foreign and Domestick*, Issue 131, 5 April 1700.

15 The humoral model was far from defunct – in a questions and answers section of the 1695 *Athenian Gazette*, a reader asking if he ought to give up tobacco was assured that Galen recommended breathing in smoke as a cure for lung ulcers (scurf, itches, chilblains and sore heels also benefitted from being rubbed with tobacco, though 'the smoak of it is bad for the brain'). The author was taking some liberties in interpretation, since tobacco wasn't present in Europe until about 1,500 years after Galen had died. ('News', *Athenian Gazette*, 12 February 1695.)

16 Dickie, S., *Cruelty and Laughter: Forgotten Comic Literature and the Unsentimental Eighteenth Century* (Chicago: University of Chicago Press, 2011).

17 Markham, G., *The English House-Wife* (London: Nicholas Okes, 1631), p. 21.

18 Meade, 'Manuscript Receipt Book of Mrs Meade, 1688–1727', Wellcome Library MS 3500, fol. 114r.

19 Swift, J., 'A Beautiful Young Nymph Going to Bed' in P. Rogers, ed., *The Complete Poems* (London: Penguin, 1983), p. 453.

20 Hargreaves, A.S., *White As Whales Bone: Dental Services in Early Modern England* (Northern Universities Press, 1998), p. 105.

21 Allen, C. (ed. L. Lindsay), *Curious Observations on the Teeth (1687)* (London: John Bale, Sons & Danielsson, Ltd, 1924), p. 13.

22 Herrick, R., 'On Glasco' in S. Romer, ed., *Robert Herrick* (New York: Faber & Faber, 2010), n.p.

23 Pepys, S., '10 October 1664', *The Diary of Samuel Pepys* [online], accessed 12 September 2023, www.pepysdiary.com.

24 *Weekly Journal, or, British Gazetteer* (London), 16 March 1717.

25 von Arni, E.G., *Justice to the Maimed Soldier: Nursing, Medical Care and Welfare for Sick and Wounded Soldiers and their Families during the English Civil Wars and Interregnum, 1642–1660* (Aldershot: Routledge, 2001), p. 185.

26 *Kingdomes Weekly Post* (London), 2 December 1645.

27 von Arni, *Justice to the Maimed Soldier*, p. 185.

28 Sewers, J., 'Advertisement' (London: John Cluer, 1710), Wellcome Library EPH 528:3.

29 Elmer, P., 'Greatrakes, Valentine [nicknamed the Stroker] (1629–1683), faith healer', *Oxford Dictionary of National Biography* [online], accessed 16 September 2024, https://www.oxforddnb.com/view/10.1093/ref:odnb/9780198614128.001.0001/odnb-9780198614128-e-11367.

30 Greatrakes, V., *A Brief Account of Mr. Valentine Greatraks, and Divers of the Strange Cures by Him Lately Performed* (London: J. Starkey, 1666), p. 22.

31 Greatrakes, *A Brief Account*, pp. 22–3.

32 Greatrakes, *A Brief Account*, p. 25.

33 Beaumont, D., 'Greatrakes, Valentine', *Dictionary of Irish Biography* [online], accessed 17 May 2024, https://www.dib.ie/biography/greatrakes-valentine-a3600.

34 Greatrakes, *A Brief Account*, p. 38–9.

35 Elmer, P., *The Miraculous Conformist: Valentine Greatrakes, the Body Politic, and the Politics of Healing in Restoration Britain* (Oxford: Oxford University Press, 2012), p. 51.

36 Greatrakes, *A Brief Account*, pp. 46–7.

37 Greatrakes, *A Brief Account*, pp. 47–8.

38 Greatrakes, *A Brief Account*, pp. 46–7.

39 Greatrakes, *A Brief Account*, p. 71.

40 Lloyd, D., *Wonders No Miracles, or, Mr. Valentine Greatrates Gift of Healing Examined* (London: Sam Speed, 1666), p. 19.

41 Lloyd, *Wonders No Miracles*, p. 21.

42 Lloyd, *Wonders No Miracles*, pp. 44–5.

43 Stubbe, H., *The Miraculous Conformist, or, An Account of Severall Marvailous Cures Performed by the Stroking of the Hands of Mr. Valentine Greatarick* (Oxford: H. Hall, 1666).

44 Elmer, *The Miraculous Conformist*, pp. 88–96.

45 Wilmot, J., 'A Satyr on Charles II' in R. DeMaria Jr., ed., *British Literature 1640–1789*, pp. 391–2.

46 *The Miserable Mountebank* (London: J. Deacon, 1685–8), *English Broadside Ballads Archive* [online], accessed 5 May 2021, https://ebba.english.ucsb.edu/ballad/33197.

Care-takers and criminals: hospitals and the welfare state

1 *London Journal (1720)* (London), Issue 131, 27 January 1722, p. 2.

2 The exception was St Benet Hulme Abbey in Norfolk, which was probably spared because its monks were famously ascetic in their habits. However, by 1545, it too had collapsed, having become mired in debt.

3 Shakespeare, *Henry IV part 2*, 1.2.250–6.

4 Paré, A. (trans. by J. Pallister), *On Monsters and Marvels* (Chicago: University of Chicago Press, 1995), p. 75.

5 Dekker, T., *O per Se O. Or A New Cryer of Lanthorne and Candle-Light Being an Addition, or Lengthening, of the Bell-Mans Second Night-Walke. In Which, Are Discouered Those Villanies, Which the Bell-Man (Because Hee Went i'th Darke) Could Not See* (London: Thomas Snodham, 1616), fol. B2v.

6 Dekker, *O per Se O*, fol. M4r.

7 *Deeds against Nature, and Monsters by Kinde Tryed at the Goale Deliuerie of Newgate, at the Sessions in the Old Bayly, the 18. and 19. of Iuly Last, 1614. the One of a London Cripple Named Iohn Arthur, That to Hide His Shame and Lust, Strangled His Betrothed Wife. The Other of a Lasciuious Young Damsell Named Martha Scambler, Which Made Away the Fru[i]t of Her Own Womb, That the World Might Not See the Seed of Her Owne Shame* (London: G. Eld., 1614), fols. A2r–v.

8 *Deeds against Nature*, fol. A3r.

9 Slack, P.A., 'Vagrants and Vagrancy in England, 1598–1664', *The Economic History Review*, 1974, 27.3, pp. 360–79, esp. p. 361.

10 Slack, 'Vagrants and Vagrancy', pp. 366–8.

11 This act was initially for twenty years, but was made permanent in 1623.

12 Fessler, A., 'Skin Diseases in 17th and 18th Century Lancashire Local History Documents', *Bulletin of the History of Medicine*, 1953, 27.5, pp. 414–9, esp. p. 415.

13 Woodall, J., 'A Treatise of Gangraena', appended to *The surgeons mate, or, Military & domestique surgery: discovering faithfully & plainly [the] method and order of [the] surgeons chest, [the] uses of the instruments, the vertues and operations of [the] medicines, with [the] exact cures of wounds made by gun-shott; and otherwise* (London: Rob, Young, 1639), p. 381.

14 Woodall, 'A Treatise of Gangraena', p. 387.

15 Woodall, 'A Treatise of Gangraena', pp. 389–90.

16 Woodall, 'A Treatise of Gangraena', pp. 389–90.

17 Woodall, 'A Treatise of Gangraena', pp. 394–5.

18 Arnold, C., *Bedlam: London and Its Mad* (London: Simon & Schuster, 2009), p. 22.

19 Drayton, M. (ed. C. Brett), *Minor Poems of Michael Drayton* (Oxford: Clarendon Press, 1907), p. 59.

20 Arnold, *Bedlam*, p. 44.

21 Andrews, J., A. Briggs, K. Waddington, R. Porter and P. Tucker, *The History of Bethlem* (London: Routledge, 1997), p. 120.

22 Andrews et al., *The History of Bethlem*, pp. 124–5.

23 Chambers, P., *Bedlam: London's Hospital for the Mad* (Shepperton: Ian Allen Publishing, 2009), p. 31.

24 Crooke, H., *Mikrokosmographia a description of the body of man. Together with the controversies thereto belonging* (London: William Jaggard, 1616).

25 Arnold, *Bedlam*, p. 62.

26 'Minute Book of Bridewell Governors Committee, March 1627–May 1634', Bethlem Museum of the Mind MS BCB-07, fol. 214.

27 Arnold, *Bedlam*, p. 62.

28 *Mercurius Aulicus* (Oxford), 10 November 1644, pp. 1,254–5.

29 *Mercurius Britanicus Communicating the Affaires of Great Britaine* (London), Issue 62, 16–23 December 1644, fol. Rrr3v.

30 von Arni, *Justice to the Maimed Soldier*, pp. 144, 147.
31 McIntosh, M.K., 'Negligence, greed and the operation of English charities, 1350–1603', *Continuity and Change*, 2012, 27.1, pp. 53–81.
32 Carlton, C., *This Seat of Mars: War and the British Isles, 1485–1746* (New Haven: Yale University Press, 2011), p. 223.
33 *A True Report of the Great Costs and Charges of the Foure Hospitals in the City of London* (London: s.n, 1645).
34 England and Wales Parliament, House of Commons, *Die Sabbati*, 24 December 1643, 'An order for a charitable contribution for the relief of maymed souldiers' (London: 1642); England and Wales Parliament, House of Commons, '17 Junii, 1643. It is this day ordered by the Commons House of Parliament, that the high-constables of the severall hundreds in the counties of Berks, Buckingham, Middlesex, and Surrey, in whose divisions any sick and maymed souldiers of the Parliaments army are or shall be billeted...' (London: 1643).
35 *Mercurius Britanicus Communicating the Affaires of Great Britaine* (London), Issue 62, 16–23 December 1644, fol. Rrr3v.
36 Rosworme, J., *Good Service Hitherto Ill Rewarded* (London: 1649), p. 27.
37 Rosworme, *Good Service*, p. 28.
38 von Arni, *Justice to the Maimed Soldier*, p. 117.
39 von Arni, *Justice to the Maimed Soldier*, p. 181.
40 Ward, *Notebook... Vol. 8*, fols. 151r, 150r.
41 O'Donoghue, E.G., *The Story of Bethlehem Hospital from Its Foundation in 1247* (Ulan Press, 2012), pp. 182–3.
42 Andrews et al., *The History of Bethlem*, p. 204.
43 Andrews et al., *The History of Bethlem*, p. 208.
44 Stevenson, C., 'The Architecture of Bethlem at Moorfields' in Andrews et al., *History of Bethlem*, pp. 230–60, esp. p. 233.
45 *Bethlehems Beauty, Londons Charity, and the Cities Glory. A Panegyrical Poem on That Magnificent Structure, Lately Erected in Moorfields, Vulgarly Called New Bedlam. Humbly Addrest to the Honourable Master, Governours, and Other Noble Benefactors of That Splendid and Most Useful Hospital.* (London: Thomas Sear [Sare], 1676).
46 Tryon, T., *A Treatise of Dreams and Visions* (London: 1689), pp. 91–2.

47 Gee, J., *The Trade and Navigation of Great-Britain Considered* (London: Sam Buckley, 1729), p. 42.

Remaking the World: Experimenters

1 The Royal Society, *Journal Book of the Royal Society Volume 3, minutes of meetings 1666 – 1668*, JBO/3, p. 100.

2 Pepys, S., '26 April 1667', *The Diary of Samuel Pepys* [online], accessed 12 September 2023, www.pepysdiary.com.

3 Pepys, '30 May 1667'.

4 Moore Smith, G.C., ed., *The Letters of Dorothy Osborne to William Temple* (Oxford: Clarendon Press, 1928), p. 41

5 Cavendish was virtually forgotten about for years, until Virginia Woolf described her in *The Common Reader*: 'There is something noble and Quixotic and high-spirited, as well as crack-brained and bird-witted, about her. Her simplicity is so open; her intelligence so active; her sympathy with fairies and animals so true and tender. She has the freakishness of an elf, the irresponsibility of some nonhuman creature, its heartlessness, and its charm.' V. Woolf, *The Common Reader: First Series* (New York: Brace & World, 1953), p. 78.

6 Ward, *Notebook... Vol. 12*, fols. 850–9; Ward, J. *Notebook of John Ward, Vol. 5*, Folger Shakespeare Library [online], accessed 5 September 2016, https://digitalcollections.folger.edu/bib335073-486298, fol. 1,123.

7 Gnudi, M.T. and J.P. Webster, *The Life and Times of Gaspare Tagliacozzi, Surgeon of Bologna 1545–1599: With a Documented Study of the Scientific and Cultural Life of Bologna in the Sixteenth Century* (New York: H. Reichner, 1950), p. 109 (quoting from *Opuscula Elisii Calentii Poetae*, Rome, 1503. Gnudi and Webster's translation).

8 Discerning the passage of an idea so long ago is always tricky, but the work of the Roman scholar Aulus Cornelius Celsus (c. 25 BCE–c. 50 CE) contains some instructions for repairing maimed lips, noses and ears which historians believe owe their origins to Sushruta's work. Sushruta's work also likely influenced the Greek physician Paulus Aegineta (c. 625–90 CE) and the Persian Avicenna (Ibn Sīnā,

c. 980–1037 CE), both of whom travelled widely across the vast expanse of the Byzantine Empire, writing works which became staples of medieval and early Renaissance physic.

9 Gnudi and Webster, *The Life and Times of Gaspare Tagliacozzi*, p. 123.

10 Tagliacozzi, G. (trans. by J.H. Thomas, ed. R. Goldwyn), *De Curtorum Chirurgia per Institionem (1597)* (New York: Gryphon Edition, 1996), p. 167.

11 Tagliacozzi, *De Curtorum*, p. 184.

12 Tagliacozzi, *De Curtorum*, p. 111.

13 'Extract from a letter to Hieronymus Mercurialis, February 10, 1586 (Frankfurt, 1587)' in Gnudi and Webster, *The Life and Times of Gaspare Tagliacozzi*, p. 139.

14 Robinson, H.W., 'An Unpublished Letter of Dr Seth Ward Relating to the Early Meetings of the Oxford Philosophical Society', *Notes and Records of the Royal Society of London*, 1949, 7.1, pp. 68–70.

15 Purver, M., *The Royal Society: Concept and Creation* (London: Routledge, 1967), pp. 86–93.

16 'Journal Book, vol II, p.179 (meeting of 18 January 1665)', cited in Purver, *The Royal Society*, p. 84.

17 Donovan, A.J., 'Richard Lower, M.D., Physician and Surgeon (1631–1691)', *World Journal of Surgery*, 2004, 28.9, pp. 938–45, esp. p. 938.

18 'Peter Lowe, letter to Robert Boyle', cited in Donovan, 'Richard Lower', p. 942.

19 Ward, *Notebook… Vol. 11*, fols. 620–2.

20 Lower, R. and R. Boyle, 'The method observed in transfusing the bloud out of one live animal into another: and how this experiment is like to be improved. Some considerations concerning the same', *Philosophical Transactions of the Royal Society of London*, 1666, 1.20, pp. 353–8.

21 Bayfield, R., *Tractatus de Tumoribus Praeter Naturam, or, A Treatise of Preternatural Tumors* (1662), p. 192.

22 Boyle, R., 'Tryals proposed by Mr. Boyle to Dr. Lower, to be made by him, for the improvement of Transfusing Blood out of one live Animal into another', *Philosophical Transactions of the Royal Society of London*, (1666), pp. 385–98.

23 Tucker, H., *Blood Work: A Tale of Medicine and Murder in the Scientific Revolution* (London: W.W. Norton, 2012), pp. 142–3.

24 Pepys, S., '21 November 1667', *The Diary of Samuel Pepys* [online], accessed 12 September 2023, www.pepysdiary.com.

25 Pepys, '20 November 1667'.

26 Though untreated blood would likely cause a deadly haemolytic reaction, in recent years, scientists have been working toward using genetically engineered pigs to supply red blood cells for human transfusion.

27 'An extract of a letter, written by J. Denis, doctor of physick, and professor of philosophy and the mathematicks at Paris, touching a late cure of an inveterate phrensy by the transfusion of bloud', *Philosophical Transactions of the Royal Society of London*, 1668, 2.32, pp. 617–24, esp. p. 622.

28 Acton, G., *Physical Reflections upon a Letter Written by J. Denis, Professor of Philosophy and Mathematicks, to Monsieur De Montmor, Counsellor to the French King, and Master of Requests Concerning a New Way of Curing Sundry Diseases by Transfusion of Blood* (London: 1668), p. 9.

29 Stubbe, H., *A Specimen of Some Animadversions upon a Book Entituled, Plus Ultra, or, Modern Improvements of Useful Knowledge Written by Mr. Joseph Glanvill, a Member of the Royal Society* (London: S.N, 1670), p. 179.

30 Shadwell, T., *The Virtuoso a Comedy* (London: Printed by T.N. for Henry Herringman, 1676), p. 31.

31 These macabre experiments were principally done by John Hunter, famous surgeon and fellow of the Royal Society, in the 1760s and 70s. The unfortunate birds (or at least, their heads) can be seen preserved in the Hunterian Museum at the Royal College of Surgeons' headquarters in London.

32 Digby, K., *Sir Kenelm Digby. Of the Sympathetick Powder. A Discourse, in a Solemn Assembly at Montpellier. Made, in French, by Sir Kenelm Digby, Knight, 1657* (London: John Williams, 1669), p. 192.

33 Gnudi and Webster, *The Life and Times of Gaspare Tagliacozzi*, p. 284.

34 van Helmont, J.B. (trans. by W. Charleston), *A Ternary of Paradoxes. The Magnetick Cure of Wounds, nativity of tartar in wine, image of God in man* (London: 1650), pp. 13–14.

35 Pulter, H., 'To Sir W.D. upon the Unspeakable Loss of the Most Conspicuous and Chief Ornament of His Frontispiece' in *Women's Works: 1625–1650*, D. W. Foster and T. Banton, eds, First edition (New York: Wicked Good Books, 2013), pp. 160–61.

36 Butler, S., *Hudibras* (London: John Murray, 1835).

37 'An Extract of a Printed Letter, addressed to the Publisher, by M. Jean Denis, D. of Physick, and Prof. of the Mathematicks at Paris, touching the differences risen about the Transfusion of Bloud' in *Philosophical Transactions of the Royal Society of London*, 1668, 3.36, pp. 710–15, esp. p. 712

Conclusion

1 Ashcroft, M.Y., ed., 'The Papers of Sir William Chaytor of Croft (1639–1721): a list with selected transcripts', *North Yorkshire County Record Office Publications*, 1984, 33, p. 188.

Index

Page numbers in **bold** refer to illustrations.